Carbon nanotube networks and their application in flexible a.

Carbon nanotube networks and their application in flexible display technology

Von der Fakultät Informatik, Elektrotechnik und Informations-
technik der Universität Stuttgart zur Erlangung der Würde eines
Doktors der Ingenieurwissenschaften (Dr.-Ing.) genehmigte Abhandlung

Vorgelegt von

Axel Schindler

aus Balingen

Hauptberichter :	Prof. Dr.-Ing. Norbert Frühauf
Mitberichter :	Prof. Dr.-Ing. Jörg Schulze
Tag der mündlichen Prüfung :	08. Juli 2019

Institut für Großflächige Mikroelektronik
der Universität Stuttgart

2019

Bibliografische Information der Deutschen Nationalbibliothek

Die Deutsche Nationalbibliothek verzeichnet diese Publikation in der Deutschen Nationalbibliographie; detaillierte bibliographische Daten sind im Internet über http://dnb.d-nb.de abrufbar.
1. Aufl. - Göttingen: Cuvillier, 2019
Zugl.: Stuttgart, Univ., Diss., 2019

D 93

© CUVILLIER VERLAG, Göttingen 2019
Nonnenstieg 8, 37075 Göttingen
Telefon: 0551-54724-0
Telefax: 0551-54724-21
www.cuvillier.de

ISBN 978-3-7369-7131-8
eISBN 978-3-7369-6131-9

3

Contents

Abbreviations and formula symbols

Abbreviations

AFM   Atomic force microscopy

AMLCD   Active-matrix liquid crystal display

AR   Aperture ratio of an active matrix. Optically active area divided by total pixel area

BCE   back channel etch

BZ   Brillouin zone

CE   Counter electrode of a liquid crystal display

CNT   Carbon nanotube; synonymously used for SWNT if not further specified

DGU   Density gradient ultracentrifugation

FOM   Figure of merit

HiPCO   High-pressure catalytic decomposition of carbon monoxide

ITO   Indium tin oxide

LCD   Liquid crystal display

MOSFET   Metal oxide semiconductor field effect transistor

NNTB   Nearest neighbor tight binding approximation

OLED   Organic light emitting diode

PDLCD   Polymer dispersed liquid crystal display

PECVD   Plasma-enhanced chemical vapor deposition

PI   Polyimide

PIPS   Polymerization induced phase separation

r-TFT   TFT with channel in radial orientation

RIE   Reactive ion etching

SAM   Self-assembled monolayer

SEM   Scanning electron microscopy

SMU   Source-measure-unit

SWNT   Single-walled carbon nanotube

t-TFT   TFT with channel in tangential orientation

TCF	Transparent conductive film
TFT	Thin-film transistor
TN	Twisted nematic (LCD)
wt%	Weight percent

Formula symbols

$\Delta\varepsilon$	Dielectric anisotropy of a liquid crystal molecule/mixture
Δn	Birefringence of a liquid crystal molecule/mixture
ε_{2p}	Energy of the $2\,p_z$ state of the isolated carbon atom in the NNTB
γ	Carbon-carbon interaction energy in the NNTB
$\hbar$	Reduced Planck's constant
λ	Optical wavelength
μ_{device}	Effective device charge carrier mobility for a CNT-TFT
ρ_c	Specific contact resistivity $[\Omega \cdot \mathrm{cm}^2]$
σ_{DC}	Electrical direct current conductivity
σ_{op}	Optical conductivity
$\varepsilon_\parallel$	Dielectric constant parallel to $\vec{n}$
$\varepsilon_\perp$	Dielectric constant perpendicular to $\vec{n}$
φ_{TC}	Figure of merit for transparent electronic conductors
$\vec{a}_1, \vec{a}_2$	Unit vectors in the graphene/CNT honeycomb lattice
$\vec{C}_h$	Chiral vector of a SWNT
$\vec{n}$	N-director of a liquid crystal molecule
$\vec{n}_{phase}$	Average n-director of LC phase
$\vec{T}$	CNT translation vector
a_0	length of the unit vectors in the graphene/CNT honeycomb lattice (2.461 Å)
a_{C-C}	distance between two carbon atoms in the graphene/CNT honeycomb lattice (1.42 Å)
C_h	Length of chiral vector and circumference of CNT
C_{GD}	Gate-drain capacity of a TFT caused by the overlap of gate and drain contacts
C_{LC}	Capacitance of a liquid crystal pixel
C_s	Storage capacitor in each AMLCD pixel parallel to C_{LC}
d	Thickness of a layer
d_{diel}	Thickness of dielectric
d_t	Diameter of a CNT
E_g	Band gap between valence and conduction band
E_{g2}	Secondary, curvature induced band gap in metallic nanotubes

f_{frame}	Frame rate of an AMLCD
g_d	$= GCD(2m+n, 2n+m)$
GCD	Greatest common divider
L_C	TFT channel length
L_t	Length of a CNT
M	Resolution-multiplicator in Monte-Carlo simulations [pixel/µm]
m	Integer describing the number of the Gooch-Tarry minima
mc	Metallic content in CNN in %
N	Number of lines in an active matrix
N	Number of runs per parameter set in Monte-Carlo simulation
N_t	Number of CNTs in channel area
$n_\perp$	Refractive index perpendicular to $\vec{n}$
$n_\parallel$	Refractive index parallel to $\vec{n}$
p_m	Percolation threshold of only the m-SWNT in a CNN
p_{mix}	Percolation threshold of a mix of s-SWNT and m-SWNT in a CNN
$R_\square$	Sheet resistance [$\Omega/\square$]
R_c	Contact resistance [Ω]
S	Subthreshold swing of a field-effect transistor
s	Overlap integral between the nearest neighbors in the NNTB
SR	Semiconducting range; Gap between TZ_m and TZ_{mix}, $SR = p_{0.5m} - p_{99.5mix}$
SR_c	Center of SR; $SR_c = \frac{p_{99.5mix} + p_{0.05m}}{2}$
T	Optical transmittance [%]
T_{TN}	Transmission trough a TN display
t_{dep}	Duration of CNT spin-coating deposition
t_{hold}	Time during which the pixel charge needs to be stored between refresh periods
t_{line}	Time available for the addressing of a single line
t_{oxide}	Thickness of the dielectric layer formed by anodic oxidation
TZ_{mix}	Transition zone of percolation probability for mix-CNNs, $TZ_{mix} = p_{99.5mix} - p_{0.5mix}$
TZ_m	Transition zone of percolation probability for m-CNNs, $TZ_m = p_{99.5m} - p_{0.5m}$
v_F	Fermi velocity
V_{max}	Maximum voltage in the anodic oxydation process
V_{CE}	Voltage applied to the counter electrode of an LCD
V_{gOFF}	Off-potential on the gate line that renders a TFT nonconductive
V_{gON}	On-potential on the gate line that renders a TFT conductive

V_{LC} Voltage applied to the liquid crystal pixel

W_C TFT channel width

Chemical symbols

Al_2O_3 Aluminum oxide

H_2O_2 Hydrogen peroxide

HNO_3 Nitric acid

n^+a-Si n-doped amorphous Silicon

NH_4OH Ammonium hydroxide

Si_3N_4 Silicon nitride

$SOCl_2$ Thionyl chloride

Ta_2O_5 Tantalum pentoxide

a-Si   amorphous Silicon

APTS   (3-Aminopropyl)triethoxysilane

LiDS   Lithium-dodecyl-sulfate

MoTa   Molybdenum Tantalum mixture

PEDOT:PSS   Poly(3,4-ethylenedioxythiophene) polystyrene sulfonate

SDS   Sodium-dodecyl-sulfate

Zusammenfassung

In dieser Arbeit wird der Einsatz von Kohlenstoffnanoröhren in der Realisierung flexibler Displays untersucht. Die Gitterstruktur der Nanoröhren besteht aus einer monomolekularen Lage aus Kohlenstoffatomen, die in einem Wabengitter angeordnet sind. Sie haben Durchmesser im Bereich von 1 nm bei Längen von 1 µm oder mehr. Aus dieser speziellen Geometrie ergibt sich eine einmalige Kombination aus elektrischen und mechanischen Eigenschaften. So gibt es sowohl metallisch leitende sowie halbleitende Kohlenstoffnanoröhren. Bei der Synthese entsteht in der Regel ein Mix der zu 1/3 aus metallischen und zu 2/3 aus halbleitenden Kohlenstoffnanoröhren besteht. Um diese nanoskopischen Moleküle grossflächig anwenden zu können werden Sie in dieser Arbeit als ungeordnete Netzwerke verwendet.

Zwei Anwendungen werden in Theorie und Experiment untersucht. Die Erzeugung transparenter, leitfähiger Schichten, die als Bildpunktelektroden Verwendung finden, sowie die Herstellung von Dünnschichttransistoren mit einem Nanoröhrennetzwerk als halbleitende Schicht. Hierfür wird das Perkolationsverhalten von metallischen und halbleitenden Nanoröhren anhand von Monte-Carlo Simulationen näher untersucht.

In der praktischen Umsetzung werden zuerst Kohlenstoffnanoröhren in Pulverform mit Hilfe von Tensiden zu einer stabilen Suspension dispergiert. Diese kann anschliessend mit einfachen und kostengünstigen Prozessen grossflächig aufgebracht werden. Für die transparenten Elektroden werden reltiv hohe Netzwerkdichten benötigt, die sich mit einem Aufsprühverfahren erzeugen lassen. Das halbleitende Netzwerk im Kanal von Dünnschichttransistoren benötigt eine feinere Kontrolle der Netzwerkdichte und wird mit einem speziellen Aufschleuderverfahren realisiert. Die Schichten werden im Anschluss per Fotolithographie und zwei möglichen Ätzschritten strukturiert.

Neben der elektrischen und optischen Charakterisierung der erzeugten Schichten werden Produktionsprozesse entwickelt, die mit der bestehenden Displaytechnologie kompatibel sind. Besonderer Wert wird dabei auf Niedertemperaturprozesse von <100 °C gelegt, um eine Produktion auf Glas- sowie flexiblen Kunstoffsubstraten gewährleisten zu können.

Für die Anwendung als transparente, leitfähige Schichten wurden komplette Displays mit Nanoröhrenelektroden hergestellt. Einfachere, segmentierte Flüssigkristallanzeigen wurden auf Glas und flexiblen Kunstoffsubstraten hergestellt. Eine auf Glas realisierte vollfarbige Aktivmatrix Flüssigkristallanzeige mit herkömmlichen Dünnschichttransistoren aus amorphem Silizium und einer Auflösung von 320xRGBx240 Bildpunkten und 4 Zoll Bilddiagonale demonstriert die Anwendbarkeit solcher Schichten und die Kompatibilität mit standardisierten Industrieprozessen. Ausgehend vom Substrat konnten die Anzeigen komplett im Rahmen dieser Arbeit hergestellt werden. Eine organische lichtemittierende Diode (OLED) mit einer Kohlenstoffnanoröhren-Anode konnte ebenfalls realisiert werden.

Es wird gezeigt wie es möglich ist, trotz des Anteils an metallischen Nanoröhren halbleitende Kanäle in Dünnschichttransistoren zu erzeugen. Mit dem stetigen Fortschritt von Aufbereitungsverfahren sind nun auch Nanoröhren mit nahezu ausschliesslich halbleitendem Anteil verfügbar. Dies führt zu einer Verbesserung der Transistorleistung. Es wird ausserdem beschrieben wie durch eine Linearbewegung der Flüssigkeitszuführung während des Schleuderprozesses die Homogenität der abgeschiedenen Schicht und somit der Transistoreigenschaften verbessert werden konnte. Der Einfluss der Netzwerktopografie auf das Transistorverhalten wird näher analysiert.

Abstract

This work describes the use of carbon nanotubes in the realization of flexible displays. Such nanotubes are hollow fibers with diameters in the range of 1 nm and lengths of 1 μm or more. They are built of a monoatomic lattice of carbon atoms, organized in a honeycomb lattice. This special configuration results in a unique combination of electrical, optical and mechanical properties. They can, for example, behave either like a metal or like a semiconductor with only slight changes in the orientation of the honeycomb lattice in reference to the nanotube circumference. As-synthesized nanotube powders typically contain a mix of 1/3 metallic and 2/3 semiconducting nanotubes. For using these nanoscopic molecules on large surfaces, they are implemented as a randomly oriented network.

Two distinct applications of carbon nanotubes in display technology are discussed in theory and experiment. The realization of pixel electrodes in the form of transparent conductive films as well as thin-film transistors that use carbon nanotube networks as the semiconducting channel. Monte-Carlo simulations are used to examine more closely the percolation probabilities of metallic and semiconducting nanotubes.

The description of the experimental part of this work starts with the methods to create stable surfactant suspensions from carbon nanotubes powders. The presentation of simple and cost-effective deposition processes for the creation of random networks from the liquid phase is followed by patterning procedures of the created films. Besides results from electrical and optical characterization, the realization of production processes, compatible with standard display technology is presented. High importance is put on a low temperature process chain for staying compatible with glass and flexible plastic substrates.

The realization of working displays is presented as proof of suitability of transparent conductive films made of carbon nanotubes. Simple segmented liquid crystal displays were realized on glass and flexible plastic substrates. The realization of a full-color active-matrix liquid crystal display with a resolution of 320xRGBx240 and 4 inch diagonal based on amorphous silicon thin-film transistors serves as proof for the compatibility of this new material with standard industry processes. All displays could be produced in the scope of this research project, starting from the bare substrate. As a further result, an organic light emitting diode (OLED) with a carbon nanotube network anode is also presented.

Functional thin-film transistors could be realized despite the metallic content in the nanotube network. Further improvements were gained by the use of nowadays available highly semiconducting nanotube feedstock and an optimized deposition process. The dependence of the TFT performance on the network density for both, mixed and highly semiconducting nanotubes, is discussed at the end of the document.

1. Introduction

The realization of flexible displays is discussed and actively pursued since many years. The main arguments are a more versatile form factor permitting new products, big foldable or rollable screens in otherwise small mobile devices, ruggedness and general technology progress. The step from the formerly glass-based flat panel display technology to real flexible devices turned however out to be quite difficult. Two significant hurdles are for example the mostly brittle materials like oxides or nitrides and process temperatures that are not compatible with flexible and transparent plastic substrates. While the existing technology can be optimized to some degree to approach the targeted goal with established materials and processes, some unbreachable limits are to be expected. New materials are therefore investigated that can fulfill the mechanical demands, exhibit a high functional performance and can be processed at low temperatures. In this thesis, carbon nanotubes in the form of randomly oriented networks are investigated and applied as transparent conductive film for pixel electrodes and as semiconducting film in thin-film transistors. Carbon nanotubes are thin, hollow fibers, formed of a monolayer honeycomb carbon lattice. This unique structure leads to some remarkable properties.

The history of carbon nanotubes (CNT) is still quite young. The discovery of the C_{60} bucky balls by Kroto, Curl and Smalley in 1985, that was later awarded with a Nobel price in chemistry, created a lot of interest in carbon nanomaterials [74]. The synthesis of this new species was however not yet well controlled. It took until 1990 that Krätschmer and Huffman published the synthesis of C_{60} molecules with a relatively simple arc-discharge reactor that uses graphite electrodes [75]. In 1991 Sumio Iijima first reported the discovery of so-called multi-walled carbon nanotubes (MWNT) that he synthesized with a similar reactor as the one used by Krätschmer [57]. The nanotubes were discovered in the carbon soot with a high resolution transmission electron microscope. Finally in 1993 two research groups were able to simultaneously present the synthesis and discovery of the more fundamental single-walled carbon nanotubes (SWNT); Iijima with his colleague Ichihashi [58] as well Bethune et.al. [10]. It was the addition of metal nanoparticles, acting as catalyst during the arc-discharge process, that led to synthesis of single-walled carbon nanotubes with narrow diameter spread in the range of 1 nm to 2 nm. Mechanical and electrical properties, that were partly already postulated by theoretical calculations before discovery of carbon nanotubes, could now be confirmed by experiment. The prediction that 1/3 of all possible SWNT geometries will behave like a 1-dimensional metal and 2/3 like a 1-dimensional semiconductor for example was already postulated in 1992 [110]. Sumio Iijima had paved the way to intense research of a new and fascinating form of carbon. CNTs already existed however long before their discovery. They could for example be discovered in an old Damascus saber from the seventeenth century [103].

It is the combination of many remarkable intrinsic properties of SWNTs that caused an interest for a huge diversity of applications, reaching from compound materials over H_2 storage, AFM tips, transistors and sensors to quantum computers, to name only a few. Some of the properties that make SWNTs interesting for the applications discussed in this work are room temperature field-effect mobilities exceeding $100\,000\,cm^2/(V\,s)$ [28], the capability to carry currents with a density up to $10^9\,A/cm^2$ combined with a tensile strength and a Young's modulus exceeding any known material plus, at the same time high mechanical flexibility [1].

Extraordinary intrinsic properties of nanomaterials are however only one side of the medal. The crux lies in exploiting these characteristics in macroscopic systems. The highly flexible tubes of only 1 nm diameter and lengths going up to micrometers or more cannot be handled individually by some kind of nano-tweezers. Locating individual tubes on a wafer by means of atomic force microscopy and defining metal contacts with ion beam lithography is a viable way for fundamental research. Production processes need however faster and better controlled manners to place and contact the nanotubes. Fundamental research further revealed, that tunneling barriers to the outside world largely limit the extrinsic performance.

In this thesis carbon nanotubes are used in the form of randomly oriented networks. These can be deposited on large surfaces with simple and cost-effective deposition processes from liquid phase. The two application fields of interest are transparent conductive films, as well as the realization of thin-film transistors with SWNT networks as semiconducting layer. At the beginning of this research project, initial results for both applications were already presented by academic research. The main focus of this work is to elaborate to which degree the CNT networks can be applied to flexible display technology and to develop production processes that can be integrated into standard liquid crystal display (LCD) technology. CNT-based transparent conductive films are integrated into rigid glass and flexible plastic based segmented liquid crystal displays as well as glass based active-matrix displays. CNT thin-film transistors realized on glass and plastic substrates are evaluated on a single device basis, which is a precursor for being able to realize active-matrices or logic circuits later on.

This thesis is structured in the following way. First, the necessary theory to help understanding SWNT physics and the demands on the application side is explained. The formation of nanotube networks is then investigated by conducting Monte-Carlo simulations. In the experimental part, the preparation of dispersions from nanotube feedstock, the tested deposition methods as well as patterning of the realized layers is discussed. Results and fabricated demonstrators are then presented in two final chapters. The first chapter discusses the realization of transparent conductive films, while results of thin-film transistor fabrication and characterization are given in the second chapter. A summary concludes the main aspects.

2. Theory

2.1. Carbon Nanotubes

This thesis investigates the use of Carbon Nanotubes (CNT) in display applications. Understanding their structure and electronic properties is important for being able to use them in macroscopic systems. In this chapter the structure and electronic properties of CNTs are described. In both cases a single atomic layer of graphite, the so-called graphene is first discussed since carbon nanotubes are basically rolled up graphene ribbons. This holds up to the point that the electronic properties can be derived from the graphene band structure.

2.1.1. Structure of graphene and single-walled carbon nanotubes

A carbon atom has 6 electrons which according to the nuclear shell model have a configuration as shown in fig. 2.1. For chemical bonding, only the upper most energy levels 2s and 2p play a role. In graphene and carbon nanotubes the carbon atoms are sp^2 hybridized. This means that the 2s orbital and the $2p_x$ and $2p_y$ orbitals form three similar sp^2 orbitals that are all occupied by one electron. These sp^2 orbitals lie in one plane with 120° between each coil. The $2p_z$ orbital keeps the 4th electron and forms a double-coil perpendicular to the sp^2 plane (see fig. 2.2a).

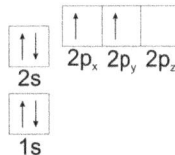

Figure 2.1.: Electron configuration of carbon according to the nuclear shell model.

When the sp^2 orbitals of two carbon atoms overlap they form a strong σ-bond. The p_z orbitals overlap as well and form a so-called π-bond with two clouds of electron probability above and below the σ-bond. The result is a so-called double bond. When several carbon atoms are connected in this way a hexagonal or honeycomb structure is formed. A single hexagon consisting of 6 carbon atoms and the remaining open sp^2 orbitals saturated with hydrogen atoms is called benzene (see fig. 2.3a). In this hexagon the π-bonds are not localized between distinct carbon atom pairs but rather form a delocalized π electron system in which the

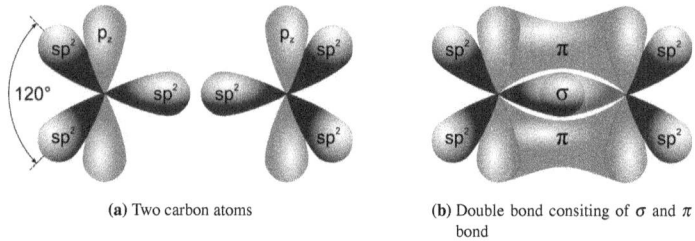

(a) Two carbon atoms

(b) Double bond consiting of σ and π bond

Figure 2.2.: Orbital configuration of and bonding between sp^2 hybridized carbon atoms.

electrons can move freely without barriers. This is indicated by the clouds above and below the molecule in fig. 2.3a.

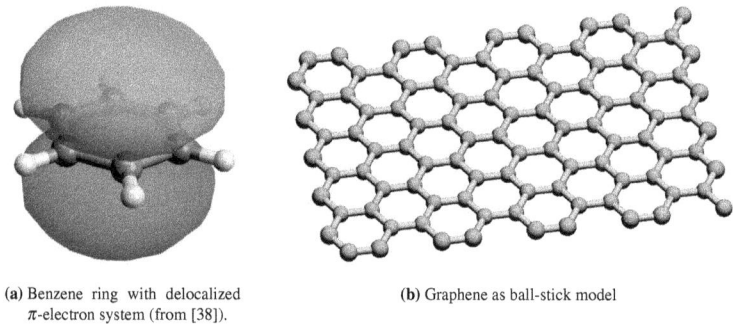

(a) Benzene ring with delocalized π-electron system (from [38]).

(b) Graphene as ball-stick model

Figure 2.3.: Hexagonal carbon molecules

A sheet of graphene results if a large number of sp^2 hybridized carbon atoms are connected. This is a monolayer crystal of hexagonally bonded carbon (see fig. 2.3b). Also in graphene the delocalized π electron system forms, giving good electrical conduction along the sheet. Graphite consists of a large number of stacked graphene sheets. Interaction between the sheets is rather weak, leading to e.g. a high anisotropy of electrical conductivity in graphite.

A single-walled carbon nanotube (SWNT) can be imagined as a rolled up finite sheet of graphene. Due to the high symmetrie of the hexagonal lattice this can be done in various manners. An unambiguous nomenclature was defined by Dresselhaus et.al. [26] (see fig. 2.4).

First the unit vectors of the honeycomb lattice $\vec{a}_1$ and $\vec{a}_2$ are defined which form an angle of 60°. Their length is

$$|\vec{a}_1| = |\vec{a}_2| = a_0 = \sqrt{3}a_{C-C} = 2.46 \,\text{Å} \tag{2.1}$$

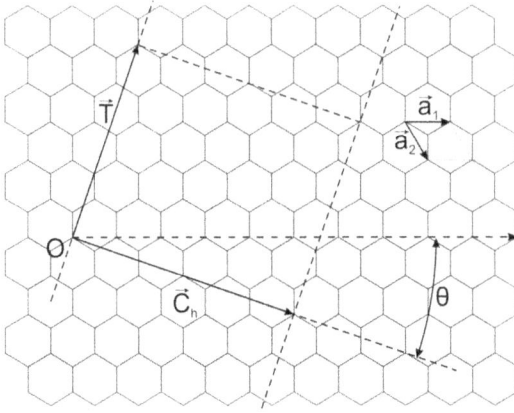

Figure 2.4.: Definition of carbon nanotube nomenclature in the hexagonal graphene lattice; Example shows graphene lattice with chirality vector $\vec{C}_h$, chiral angle θ and translation vector $\vec{T}$ of a (4,2) chiral SWNT.

where the distance between two carbon atoms is

$$a_{C-C} = 1.42\,\text{Å} \tag{2.2}$$

The chirality vector

$$\vec{C}_h = n\vec{a}_1 + m\vec{a}_2 \equiv (n,m); \qquad \{n,m\} \in \mathbb{N}; \qquad 0 \le |m| \le n \tag{2.3}$$

describes the circumference of the nanotube. It connects two crystallographic identical positions that are superimposed when the sheet is theoretically rolled up to form the tube. The 2-tuple (n,m) non-ambiguously describes the single-walled carbon nanotube and all geometric parameters can be calculated thereof [27, p.7]. Also physical properties like e.g. the electronic behavior can be derived as will be shown in the next section. The length of the chiral vector and therefore the circumference of the nanotube can by calculated as:

$$C_h = |\vec{C}_h| = \sqrt{\vec{C}_h \cdot \vec{C}_h} = a_0 \sqrt{n^2 + nm + m^2}. \tag{2.4}$$

The angle between $\vec{C}_h$ and $\vec{a}_1$ is called the chiral angle.

$$\theta = \arctan \frac{\sqrt{3}m}{2n+m} \tag{2.5}$$

Due to the high symmetry it holds $0° \le |\theta| \le 30°$. There are three distinct classes of SWNTs depending on the value of θ - zigzag, armchair and chiral nanotubes. A list of the specifications for these three classes is given in table 2.1 while representative images are shown in fig. 2.5. The upper part of the displayed tubes indicate where the names are coming from. They describe the form of the lattice along the chiral vector.

Table 2.1.: SWNT categories and their specific properties.

Name	θ	(n,m)
zigzag	$0°$	$(n,0)$
armchair	$30°$	(n,n)
chiral	$0° < \theta < 30°$	$n \neq m;\ \ n,m \neq 0$

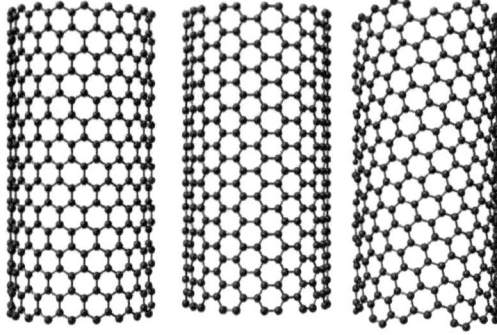

Figure 2.5.: Single walled carbon nanotubes with $d_t = 1.49$ nm, left: zigzag (19,0), center: armchair (11,11), right: chiral (16,5), red molecules indicate origin of type name.

The diameter of the tube is easily calculated from the length of the chiral vector.

$$d_t = \frac{C_h}{\pi} = \frac{a_0}{\pi} \sqrt{n^2 + nm + m^2} \tag{2.6}$$

Another characteristic parameter is the translation vector $\vec{T}$ that stands perpendicular on $\vec{C}_h$ and therefore is parallel to the cylinder axis. Both vectors start in the same origin point O. The length of $\vec{T}$ is determined by the closest intersection with a crystallographic identical atom in the lattice. It can also be described by n and m as [138, p.80]

$$\vec{T} = \left(\frac{2m+n}{g_d}, -\frac{2n+m}{g_d} \right) \tag{2.7}$$

where g_d is the greatest common divider of $2m+n$ and $2n+m$ or

$$g_d = GCD(2m+n, 2n+m). \tag{2.8}$$

It's length is derived thereof as

$$|\vec{T}| = \frac{\sqrt{3}\pi d_t}{g_d}. \tag{2.9}$$

$\vec{C}_h$ and $\vec{T}$ span up the unit cell of the SWNT which is in contrast to the rectangular form indicated in fig. 2.4 in reality a cylinder with diameter d_t and length T.

For the later discussed electronic properties of CNTs, the number of hexagons per CNT unit cell are another important information. It can be calculated by dividing the area of the CNT unit cell by the area of one hexagon which is equal with the area spanned up by the unit vectors $\vec{a}_1$ and $\vec{a}_2$ [138, p.80].

$$N = \frac{|\vec{C}_h \times \vec{T}|}{|\vec{a}_1 \times \vec{a}_2|} = \frac{2(n^2 + nm + m^2)}{g_d} = \frac{2C_h^2}{a_0^2 g_d} \tag{2.10}$$

2.1.2. Electronic properties of graphene and single walled carbon nanotubes

A basic concept in solid state physics for visualizing and understanding the band structure of crystalline materials with a regular lattice are the Brillouin zones (BZ) in reciprocal space. Constructing and interpreting the Brillouin zones for 3D lattices can become quite complex. In the case of the quasi 2D structure of graphene it is rather simple. A detailed general description of this concept with graphene as an example can be found in [138]. This section will give a short summary.

The reciprocal lattice is the Fourier transform of the direct lattice. Equivalent to the frequency domain which is the Fourier transform of the time domain. It demands a Bravais lattice in direct space as starting point. A Bravais lattice is a fundamental lattice where each lattice point "sees" exactly the same environment. As can be seen in fig. 2.6a the honeycomb lattice is by itself not a Bravais lattice. There are two kinds of sites A and B that see a different environment. It can however be converted into a Bravais lattice by defining the atoms A and B as the basis. The result is a hexagonal lattice and can be visualized by only regarding the black points (see fig. 2.6b). The lattice vectors given in reference to the cartesian coordinates are

$$\vec{a}_1 = \left(\frac{\sqrt{3}a_0}{2}, \frac{a_0}{2} \right), \qquad \vec{a}_2 = \left(\frac{\sqrt{3}a_0}{2}, \frac{a_0}{-2} \right) \tag{2.11}$$

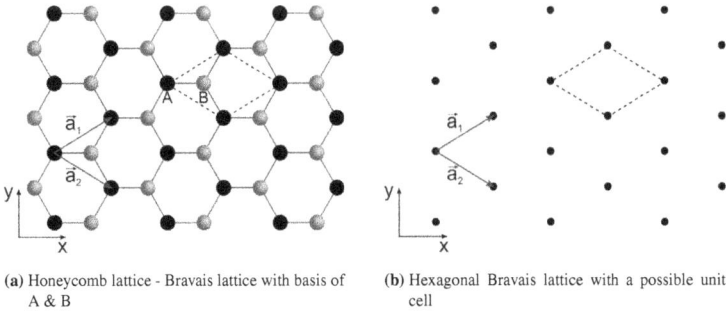

(a) Honeycomb lattice - Bravais lattice with basis of A & B

(b) Hexagonal Bravais lattice with a possible unit cell

Figure 2.6.: Conversion of honeycomb lattice to Bravais lattice with basis of two atoms.

The reciprocal lattice which is shown in fig. 2.7a is also a hexagonal Bravais lattice which is however turned by 90° in reference to the direct lattice. It's unit vectors are

$$\vec{b}_1 = \left(\frac{2\pi}{\sqrt{3}a_0}, \frac{2\pi}{a_0} \right), \qquad \vec{b}_2 = \left(\frac{2\pi}{\sqrt{3}a_0}, \frac{-2\pi}{a_0} \right). \tag{2.12}$$

The Brillouin zone can be constructed by choosing an individual lattice point and determining the mirror axes with each one of it's nearest neighbors. The area that these axes encase is the 1st Brillouin zone (see hexagon in fig. 2.7a). Further BZs could be constructed by choosing the 2nd nearest neighbors and so on. These higher order BZs only duplicate however the information that is already represented by the 1st BZ. Hence, when talking about the Brillouin zone without further notice usually the 1st one is meant. Some further common conventions: In the direct lattice, positions are referenced by the position vector $\vec{R}$ in units of length. The equivalent in the reciprocal lattice is the wave vector $\vec{K}$ in units of length^{-1}. This is why reciprocal space is alternatively called k-space.

The resulting BZ of graphene with it's high symmetry points Γ, K and M is shown in fig. 2.7b. 2D plots of the band structure are usually given along the direct connecting lines between these points. It has to be noticed that there exist six K and six M points in the graphene BZ. The K-points are often further alternately called K and K'. For the electronic properties of graphene and CNTs discussed in this chapter this differentiation is however not relevant.

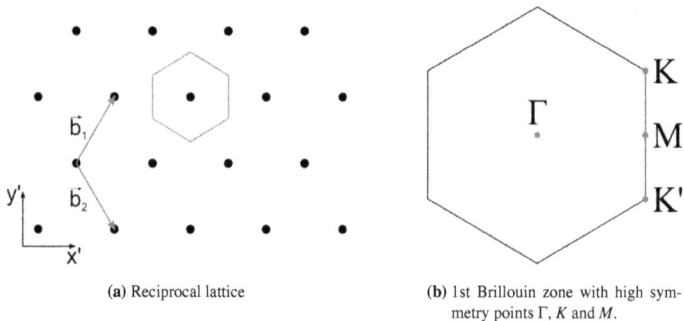

(a) Reciprocal lattice

(b) 1st Brillouin zone with high symmetry points Γ, K and M.

Figure 2.7.: Hexagonal Bravais lattice transformed to reciprocal space.

2.1.2.1. Graphene - the tight binding approximation

The band structure of a solid crystal can be either determined by doing numerical ab-initio computation based on a physical model or by searching for an analytical solution, which demands solving Schrödinger's equation

$$H\Psi(\vec{k}) = E(\vec{k})\Psi(\vec{k}), \tag{2.13}$$

where H is the Hamiltonian, $E(\vec{k})$ are the eigenvalues at wave vector $\vec{k}$ and $\Psi(\vec{k})$ are the eigenfunctions [104].

It has been shown by Saito et al. that a relatively simple analytic equation can be derived by using the so-called tight binding approximation [108]. Somehow in contrast to the above stated fact, that electrons in graphene form a free electron gas, the tight binding approximation sees the outer most valence electrons localized close to their atom cores. A continuous band structure is nevertheless received by incorporating interaction with surrounding neighbors.

A great simplification is achieved by only calculating the band structure of the π-electrons, neglecting the electrons in the sp^2 orbitals. This is possible since sp^2 and p_z orbitals do not overlap and the energy levels of the strong sigma bonds are farther away from the Fermi energy and therefore do not contribute to the electronic transport in the molecule. The second common significant simplification is to consider only a limited range of interaction with neighboring electrons. The simplest case is the so-called nearest neighbor tight binding approximation (NNTB). In reference to fig. 2.6a a type A atom interacts only with it's three neighbors of type B. The resulting energy dispersion is given in eq. (2.14).

$$\boxed{E^{\pm}(\vec{k}) = \frac{\varepsilon_{2p} \pm \gamma w(\vec{k})}{1 \pm sw(\vec{k})}} \qquad (2.14)$$

with

$$w(\vec{k}) = \sqrt{1 + 4\cos\frac{\sqrt{3}k_x a}{2}\cos\frac{k_y a}{2} + 4\cos^2\frac{k_y a}{2}} \qquad (2.15)$$

where ε_{2p} represents the energy of the $2p_z$ state of the isolated carbon atom and is usually pinned to the Fermi level ($\varepsilon_{2p} = 0\,\mathrm{eV}$), γ is the carbon-carbon interaction energy and s is the overlap integral between nearest neighbors. Since the unit cell contains two π electrons (p_z-electrons of molecules A & B), eq. (2.14) describes two energy bands. When using the positive signs the result is the bonding π-band or valence band. The negative signs yield the anti-bonding π^*-band or conduction band. The parameters ε_{2p}, γ and s are often used as fitting parameters in order to represent as close as possible the bands found by ab-initio computations or experimental data. In [104] they are set to $\varepsilon_{2p} = 0\,\mathrm{eV}$, $\gamma = 2.84\,\mathrm{eV}$ and $s = 0.07$. The resulting energy dispersion is shown in fig. 2.8.

The resulting band structure is rather peculiar. The valence and conduction bands do overlap, but only in the six K-points. Graphene is therefore considered as semi-metal or zero-bandgap semiconductor. In the following section it will be shown how this influences the electronic behavior of carbon nanotubes.

When comparing the results of the nearest neighbor tight-binding approximation with experimental data or results of ab-initio calculations it becomes obvious that for higher energies - meaning farther away from the Fermi level - there are quite significant discrepancies and the NNTB approximation cannot describe sufficiently close the band structure of graphene. The general form is however nevertheless quite good described and especially for lower energies ($E_F \pm 1\,\mathrm{eV}$) there is a good congruence. Since this is the region

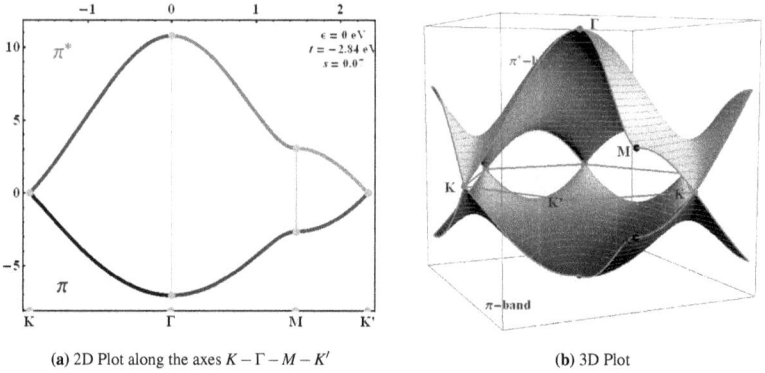

(a) 2D Plot along the axes $K - \Gamma - M - K'$ (b) 3D Plot

Figure 2.8.: Valence/π-band and conduction/π^*-band of graphene in the 1st Brillouine zone from NNTB approximation (eq. (2.14)); fitting parameters are set to $\varepsilon_{2p} = 0\,\mathrm{eV}$, $\gamma = 2.84\,\mathrm{eV}$ and $s = 0.07$ (graphics created with [33]).

that defines the electronic transport in graphene, the NNTB approximation is widely used to derive the electronic structure of SWNTs as will be described in the next section.

S. Reich et al. extended the tight binding approximation to the 3rd nearest neighbor, giving a highly increased agreement [105]. The resulting equation becomes however less handy. For the general evaluation of the SWNT electronic properties we will therefore stay with the NNTB approximation as is customary in CNT literature.

2.1.2.2. Carbon nanotubes - the zone-folding method

We already discussed the CNT unit cell in section 2.1.1. It is defined by the chiral vector $\vec{C}_h$ and the translation vector $\vec{T}$. Their counter parts in k-space are the reciprocal lattice vectors along the circumferential direction

$$\vec{K}_c = \frac{(m+2n)\vec{b}_1 + (n+2m)\vec{b}_2}{2(n^2 + nm + m^2)} \tag{2.16}$$

and along the nanotube axis

$$\vec{K}_a = \frac{m\vec{b}_1 - n\vec{b}_2}{N} \tag{2.17}$$

which together describe the Brillouin zone of the nanotube. Both are defined in reference to the unit vectors of graphene's reciprocal lattice $\vec{b}_1$ and $\vec{b}_2$. A detailed derivation of $\vec{K}_c$ and $\vec{K}_a$ can be found in [138, Ch4.4] or with slightly changed nomenclature in [108, Ch3.5].

The length of these reciprocal lattice vectors is inversely proportional to their counter parts in direct space and can be calculated to

$$|\vec{K}_c| = \frac{2\pi}{C_h} \tag{2.18}$$

and

$$|\vec{K}_a| = \frac{2\pi}{T} \tag{2.19}$$

respectively.

Not all wave vectors in the Brillouin zone defined by $\vec{K}_c \times \vec{K}_a$ are allowed however. For a nanotube with infinite length L_t ($L_t \gg T$) the component along the nanotube axis stays continuous. It is usually labeled k and when centered to the Brillouin zone (Γ-point in the graphene BZ) it represents a continuous variable in the range of

$$k = \left(-\frac{\pi}{T}, \frac{\pi}{T}\right). \tag{2.20}$$

In the circumferential direction, the wave vector q needs to fulfill quite changed boundary conditions compared to graphene since the CNT is composed of a more or less narrow ribbon of graphene lattice with joined ends. The range of q that can be derived from these periodic boundary conditions is

$$q = \frac{2\pi}{C_h} j \qquad (j = 0, 1, \ldots, N-1). \tag{2.21}$$

It is obvious that the value j introduces a distinct quantization in the direction of $\vec{K}_c$.

A summarized description of the allowed k-vectors in the CNT Brillouin zone is given by Equation (2.22)

$$\vec{k} = k\frac{\vec{K}_a}{|\vec{K}_a|} + q\frac{\vec{K}_c}{|\vec{K}_c|} \tag{2.22}$$

where $\frac{\vec{K}_a}{|\vec{K}_a|}$ and $\frac{\vec{K}_c}{|\vec{K}_c|}$ represent the unit vectors in axial and circumferential directions respectively. Substituting Equations (2.18), (2.19) and (2.21) in Equation (2.22) leads to the expression

$$\boxed{\vec{k} = k\frac{\vec{K}_a}{2\pi/T} + j\vec{K}_c \qquad \left(-\frac{\pi}{T} < k < \frac{\pi}{T}; \quad j = 0, \ldots, N-1\right)}. \tag{2.23}$$

This equation looks somehow cryptic, it can however easily be imaged. It basically describes that for each value of j there is a 1D line with a length ranging from $-\frac{\pi}{T}$ to $\frac{\pi}{T}$, i.e. the length of $\vec{K}_a$. The distance of these lines is given by $\vec{K}_c$. Often these lines are also centered to graphene's Γ-point and j is therefore varied in the range of $(-\frac{N}{2}+1 < j < \frac{N}{2})$ instead of the range given in Equations (2.21) and (2.23). The resulting lines superimposed on the graphene BZ for representatives of the three nanotube types (armchair, zigzag and chiral) are shown in fig. 2.9. The area of the CNT Brillouin zone (spanned up by the cutting lines) is identical to the area of the graphene BZ [138, Ch4.4].

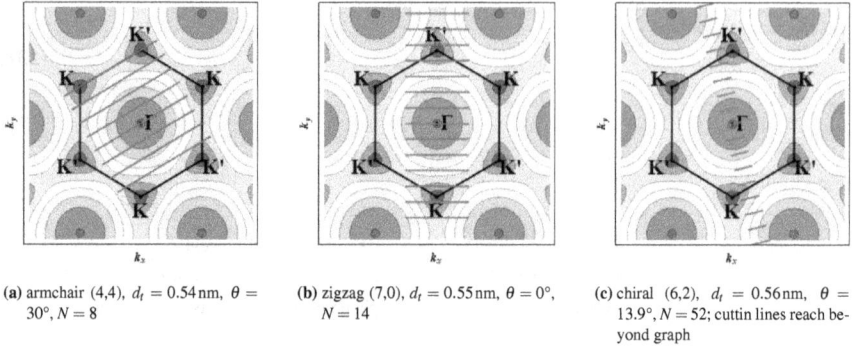

(a) armchair (4,4), $d_t = 0.54\,$nm, $\theta = 30°$, $N = 8$

(b) zigzag (7,0), $d_t = 0.55\,$nm, $\theta = 0°$, $N = 14$

(c) chiral (6,2), $d_t = 0.56\,$nm, $\theta = 13.9°$, $N = 52$; cuttin lines reach beyond graph

Figure 2.9.: Brillouin zone (red cutting lines) for three different CNTs with diameters between 0.54 nm and 0.56 nm superposed on the graphene Brillouin zone; distance and orientation of red lines defined by $\vec{K}_c$, length of red lines equals $|\vec{K}_a|$ as indicated in fig. 2.12b, θ in reference to k_y axis; cutting lines are shown in extended BZ scheme, lines can be folded into first graphene BZ resulting in N continuous lines with distance K_c that cover the complete graphene BZ; graphics created with [2].

The result of this zone folding method is that the derived set of wave vectors describe the allowed states in graphene's BZ. The 1D band structure of a given nanotube can therefore be constructed by cutting the above derived graphene 2D band structure along all N lines (hence called cutting lines) and superimposing the resulting conduction and valence bands. This is shown for the same three nanotubes in fig. 2.10.

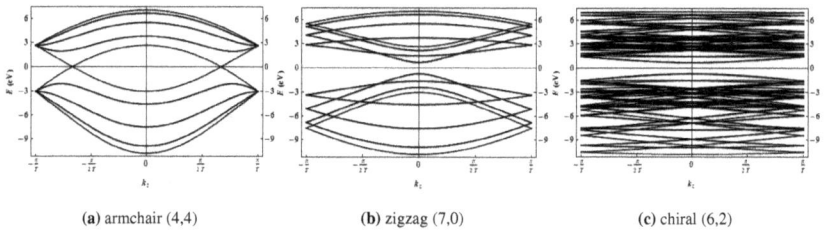

(a) armchair (4,4)

(b) zigzag (7,0)

(c) chiral (6,2)

Figure 2.10.: 1D band structure of CNTs from fig. 2.9 derived from graphene NNTB approximation with $\varepsilon = 0$, $\gamma = 2.84\,$eV and $s = 0.07$; graphics created with [2].

The peculiar band structure of graphene where the valence and conduction band are only touching exactly in the K-points has a strong effect on the electronic nature of varying nanotubes. This becomes obvious when comparing the sub-figures of fig. 2.10. While for the armchair nanotube (fig. 2.10a) valence and conduction bands cross at the Fermi level leading to a metallic tube, there is a distinct band gap for the zigzag (fig. 2.10b) and the chiral nanotube (fig. 2.10c), resulting in semiconducting tubes.

With the knowledge of the graphene band structure this becomes already evident from the cutting lines in fig. 2.9. A nanotube can only be metallic when one of the cutting lines crosses a K-point. In all other cases there is a defined band gap resulting in a semiconducting nanotube.

It can be shown that the condition for a nanotube to be metallic is that $(2n + m)$ is a multiple of 3 [108, Ch 4.1]. This is always the case for armchair (n,n) nanotubes. Zigzag and chiral nanotubes can be metallic or semiconducting. A summary of electronic type and diameter for all theoretically possible tube geometries up to $d_t \approx 1.6$ nm is given in fig. 2.11.

For a homogeneous distribution of different nanotube types this results in a ratio of 1/3 metallic CNTs to 2/3 semiconducting CNTs. When growing nanotubes, often a restricted range of diameter distributions is generated. Further, certain types of nanotubes might be favored depending on bonding energies. Both effects are depending on the used method and chosen growth conditions. The resulting nanotube powder might therefore slightly differ from the 1/3 to 2/3 ratio. Further details can be found in section 4.1.

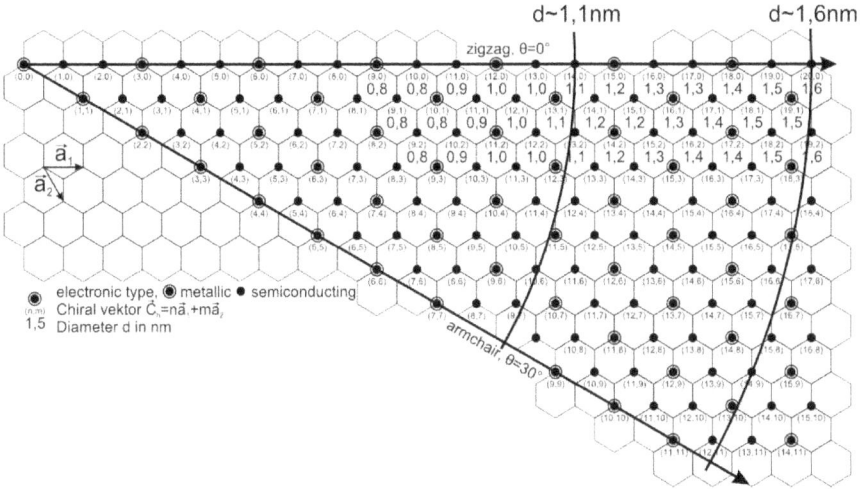

Figure 2.11.: CNT types; The zone between 1.1 nm and 1.6 nm indicates the diameter range for the most common CNT powders.

The gap in the band structure of semiconducting nanotubes E_g which is always a direct band gap can be approximated as [138, Ch 4.8]

$$E_g \approx 2\gamma \frac{a_{c-c}}{d_t} \approx 0.9 \,\text{eV}\,\text{nm}/d_t. \qquad (2.24)$$

Given the largely differing CNT BZs for similar diameter tubes (compare fig. 2.9), and the quite regular distribution of metallic and semiconducting nanotubes over the whole diameter range, the proportionality of $E_g \propto \frac{1}{d_t}$ without any influence of chirality or (n,m) is rather surprising. Both the direct band gap and the diameter dependence can be visualized by having a closer look at the graphene 2D band structure around the K-point. For energies in the range up to $\pm 0.6\,\text{eV}$ around E_F the valence and conduction band have a cone-like form and can be described by the simple linear equation

$$E(\vec{k})^{\pm} = \pm \hbar v_F |\vec{k}|; \qquad v_F = \frac{1}{\hbar}\frac{\partial E}{\partial k} \tag{2.25}$$

where $\hbar$ is the reduced Planck's constant, v_F is the Fermi velocity (velocity of electrons with energy E_F) and the K-point is at $\vec{k} = (0,0)$[138, Ch 3.6]. Again the negative and positive signs describe valence and conduction band, respectively. The form described by eq. (2.25) is also called Dirac cone and is in good approximation further valid up to the often mentioned range of $E_F \pm 1\,\text{eV}$. It is depicted in fig. 2.12a where in contrast to eq. (2.25) $\vec{k} = (0,0)$ corresponds to the Γ-Point of the graphene BZ. With the graphene band structure in mind (see fig. 2.8) obviously the minimum band gap is caused by the cutting line passing closest to the K-point. The double-cone form therefore results in a direct band gap.

While for metallic CNTs one of the cutting lines hits the K-point it can be shown that for all semiconducting nanotubes the closest cutting line passes always at a distance of $\frac{1}{3}\vec{K}_c$ (indicated by the dotted lines in fig. 2.12b) [27, 109]. As eqs. (2.6) and (2.18) reveal it holds that

$$K_c \propto C_h^{-1} \propto d_t^{-1}. \tag{2.26}$$

Hence for semiconducting CNTs the cutting line closest to the K-point moves closer to the latter with increasing diameter, resulting in a diminishing band gap due to the Dirac double-cone. A plot of the ap-

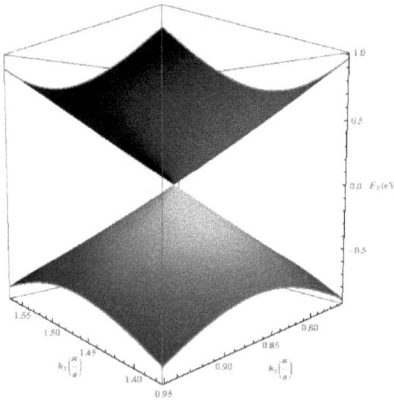

(a) Linear approximation of Dirac cone in the range of $E_F \pm$ 1 eV (graphic created with [33]).

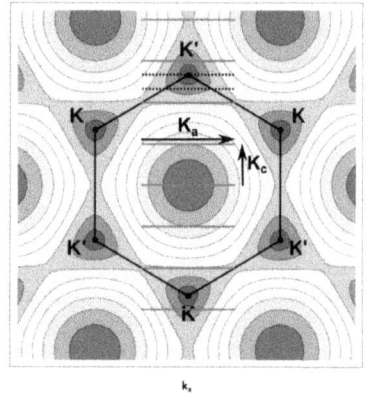

(b) Cutting lines of a semiconducting (4,2) SWNT not hitting K-point; dotted lines indicate 1/3 step between cutting lines; graphic created with [2] and further adapted.

Figure 2.12.: K-point of the graphene Brillouin zone.

proximated band gap from eq. (2.24) over the tube diameter d_t is shown in fig. 2.13 for a diameter range of $0.8\,\text{nm}$ to $2\,\text{nm}$.

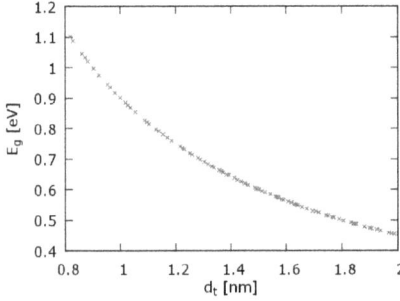

Figure 2.13.: Bandgap E_g vs. tube diameter d_t for semiconducting CNTs.

2.1.2.3. Limitations of the zone folding method

In the recent two subsections the electronic properties of CNTs were derived by first describing the band structure of graphene with the help of the nearest neighbor tight binding approximation and then cutting out the bands allowed by the quantization introduced by the circumferential boundary conditions using the zone folding method. This approach is very powerful since it allows to quickly calculate the electronic bands of any CNT by using a set of relatively simple equations. It should however not be forgotten that some simplifications were chosen that limit the accuracy and validity of the derived characteristics.

1. As was discussed in section 2.1.2.1 the NNTB approximation describes well the general form of graphene's 2D band structure but the absolute values differ quite significantly for higher energies. Often a distance of $\approx \pm 1\,\mathrm{eV}$ from the Fermi level E_F is given as acceptable energy range.

2. The zone-folding method described in section 2.1.2.2 only takes into account the circumferential boundary conditions but neglects the influence of curvature on the binding geometries and energies.

The restrictions given in the first point are sufficient for the electronic transport properties discussed in the following applications. The second point introduces a small shift in the above derived CNT Brillouin zone, resulting in a secondary band gap in metallic CNTs [66]. This secondary, curvature induced band gap is depending on the tube diameter d_t and the chiral angle θ and can be approximated to [71]

$$E_{g2} = \frac{\gamma a_0^2}{4 d_t^2} \cos 3\theta. \tag{2.27}$$

Equation (2.27) becomes zero for $\theta = 30°$ which is the case for all armchair CNTs. This is because the introduced shift is parallel to the allowed cutting lines. For all other metallic nanotubes with $\theta \neq 30°$ E_{g2} is non-vanishing but small compared to the primary band gap of the semiconducting CNTs in the relevant range down to $d_t = 1\,\mathrm{nm}$. Figure 2.14 shows values for a diameter range of 0.8 nm to 2 nm. In the practical diameter range these secondary band gaps are mostly below 30 meV. The thermal excitation of charge carriers around

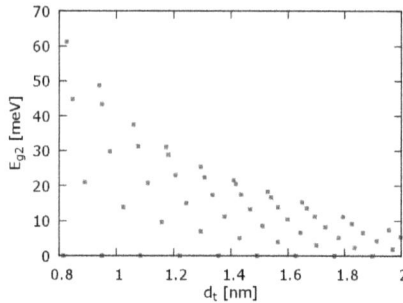

Figure 2.14.: Secondary band gap E_{g2} vs. tube diameter d_t for metallic CNTs; the points at $E_{g2} = 0\,\text{eV}$ represent armchair CNTs.

room temperature is sufficient to consider even the non-armchair metallic CNTs with curvature induced band gap as quasi-metallic.

2.1.3. Density of states in carbon nanotubes

The density of states (DOS) describes the number of available states that a charge carrier with a given energy can occupy. A plot for the number of states vs. energy for both a semiconducting and a metallic nanotube is given in fig. 2.15. The peaks that are visible in both plots are characteristic for any 1D system and are called Van Hove singularities. At these energies the number of available states drastically increases. What is further characteristic in the two plots is the band gap with no available states between $-0.5\,\text{eV}$ and $0.5\,\text{eV}$ for the semiconducting nanotube in fig. 2.15b and the comparably small but constant number of states between the first two Van Hove singularities around the Fermi Energy (E=0 eV) for the metallic nanotube in fig. 2.15a.

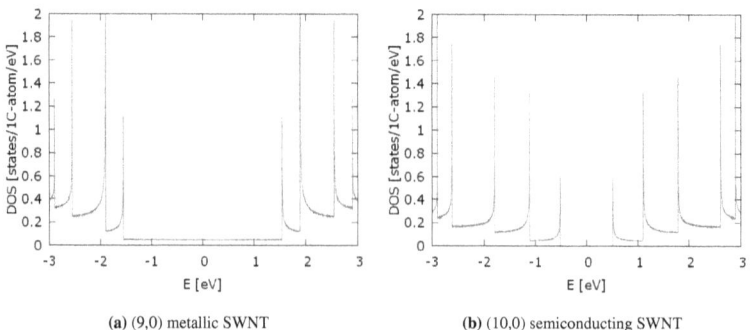

(a) (9,0) metallic SWNT

(b) (10,0) semiconducting SWNT

Figure 2.15.: Density of states for two zigzag carbon nanotubes. Data courtesy of S. Maruyama [88].

2.1.4. Carbon nanotube networks - electrical transport and switching

The remarkable physical properties of carbon nanotubes are often extracted intrinsic values of individual tubes. The confirmation of such values by experiments is often very elaborate due to the nanoscopic scale of the tubes and a lack of easy possibilities to deposit or grow nanotubes at a predefined position. As in this work the practical use in macroscopic systems is the main aspect, carbon nanotube networks (Cnns) are used as building block rather than individual tubes. Therefore aspects such as transport between tubes and between metal layers and tubes play an important role in the overall device behavior. Some paragraphs in this section are taken from [118], which the author of this thesis wrote himself.

The conduction in carbon nanotube networks was studied by several groups in theory and experiment using for example electric force microscopy [131]. A simple model to describe a CNT network (CNN) is by thinking of the individual nanotubes as straight sticks with diameter d_t and length L_t (see fig. 2.16). The aspect ratios of individual tubes can reach from $L_t/d_t = 100$ to several thousand or in extreme cases above 1,000,000. By using percolation theory, a basic understanding of the conduction through a CNN with the classical mixture of 1/3 m-SWNTs can be gained [53, 78, 125] (see also chapter 3).

In the following the network conduction will be explained for the case of a thin-film transistor with channel length L_C and width W_C and parallel source and drain electrodes on both sides. The same principles however also hold for the application as transparent conductive film. In the case of $L_C \gg L_t$, the behavior of the CNN can be tailored by adjusting its network density D [tubes/area]. Four distinct regimes can be characterized: (1) For very low values of D no conductive paths between source and drain exist, leading to an open device (see fig. 2.16a). (2) When D is just above the percolation threshold of the mixed network of s-SWNTs and m-SWNTs p_{mix}, individual current paths between source and drain exist (see fig. 2.16b). As long as D stays below the percolation threshold of only the m-SWNTs p_m all current paths are effectively semiconducting and the current flow can be controlled by the gate bias. A low off-current I_{off} can therefore be achieved. (3) For increasing densities but $D < p_m$ more current paths are created, leading to a higher conductance of the network while still being semiconducting. (4) When $D > p_m$ more and more purely metallic paths connect the source and drain contacts, leading to increasing I_{off} and smaller I_{on}/I_{off} up to a point where no gating effect is perceivable any more [81]. Further increasing D creates networks with high electrical conductivity. The optical transmission decreases at the same time. A detailed study of this electro-optical correlation will be presented in section 6.2. The influence of D on the TFT performance is discussed in section 7.4.

When using a CNN as electrical conductor or semiconductor the intertube contact resistance is an important information to understand the limitations. It was found by experiment that the junction resistance between crossed nanotubes is much larger than the intratube resistance of the SWNTs themselves [31, 127]. At junctions of two m-SWNTs or two s-SWNTs, charge carriers need to tunnel a thin barrier. Judging by the size of the contact area these junctions make relatively good tunnel contacts. In the case of a junction between metallic and semiconducting SWNTs, there exists an additional Schottky barrier which is increasing the contact resistance by two orders of magnitude [31]. This Schottky barrier does not only need to be over-

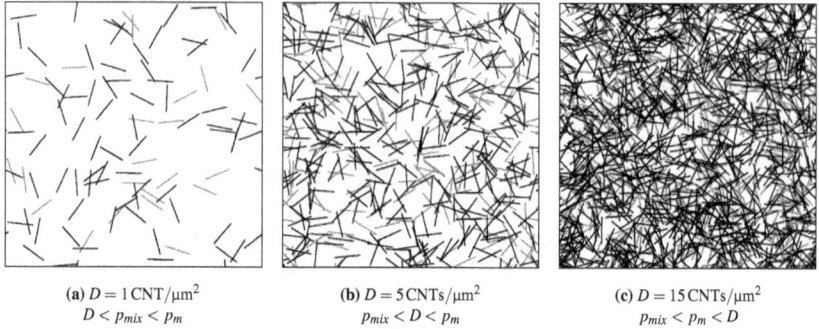

(a) $D = 1\,CNT/\mu m^2$
$D < p_{mix} < p_m$

(b) $D = 5\,CNTs/\mu m^2$
$p_{mix} < D < p_m$

(c) $D = 15\,CNTs/\mu m^2$
$p_{mix} < p_m < D$

Figure 2.16.: $10\,\mu m \times 10\,\mu m$ patches of random CNNs of 1/3 m-SWNT (red sticks) and 2/3 s-SWNT (black sticks) for different network densities D, SWNT length $L_t = 1\,\mu m$.

come for carriers going from one tube to the other. The originating space charge region at the intersection also limits carrier transport along the s-SWNT.

Switching in CNT-TFTs is therefore not only dominated by the gate-induced modulation of charge carrier density in s-SWNTs. The gate potential also modulates Schottky barriers between metal contact and CNN as well as Schottky barriers in the network itself. In classical MOSFETs on the other side, switching is caused by modulation of the carrier density in the channel. Other than in classical Schottky barrier devices, the subthreshold swing S (see section 2.3.3.4) in single CNT transistors can however come close to the theoretical limit of $60\,mV/decade$ [62]. CNN-TFTs exhibit however often higher values for S that can go beyond $1\,V/decade$.

The increased conduction resistance for carrier transfer between tubes limits the effective charge carrier mobility μ_{eff} for network TFTs by about 2 orders of magnitude compared to single tube devices [140]. The limitations caused by the intertube resistance can be reduced by achieving more parallel alignment [73] or by enrichment in s-SWNTs [131]. In the case of $L_C \gg L_t$ a certain degree of misalignment is however necessary to get a well connected oriented network. While a highly aligned deposition of nanotubes is not possible with the deposition methods used in this thesis, mixtures with enriched semiconducting content are tested.

The quantum physical barriers at tube junctions are not the only effects limiting carrier mobility in a CNN. The topography and purity of the network is further influencing device performance. CNT length, defects, bundling, impurities and contamination that might be introduced during synthesis or later processing will limit the device performance. In particular the surfactant used for dispersing nanotubes in a liquid phase could increase the tunneling barrier between individual tubes or between CNTs and metal contacts. Complete removal of the residual surfactant is therefore an important factor for performance improvement. Geng et.al. reported that surfactant removal by nitric acid treatment can increase electrical CNN conductivity by almost a factor of 3 [34]. Blackburn et.al. confirmed surfactant removal and conductivity improvement by HNO_3 and $SOCl_2$. He could however show that the effect is completely reversible [12]. Based on Raman spectroscopy they come to the conclusion that both chemicals act as redox dopants, shifting the fermi level towards the valence band and increasing the free carrier density of holes. The s-SWNTs are therefore p-doped. The dopants further lower the tunnel barriers, improving the conductance through the nanotube junctions. O_2 was also reported to act as redox dopant [21]. By calculations Blackburn et.al. show that a high degree of doping by acid or O_2 can shift the Fermi level in s-SWNTs into the valence band where due to the van-Hove singularities the E_F will end up at higher density of states than in m-SWNTs (compare fig. 2.15). They postulate that highly doped CNNs are therefore dominated by conduction in s-SWNTs. A purely semiconducting CNN with sufficient doping will therefore have a higher conductivity than a purely metallic CNN. As the dopants are volatile, metallic CNNs will however be the more stable solution. As a result of p-type doping by O_2 and metal-CNT contact CNT-TFTs, both single tube and CNN devices, behave as p-type devices under ambient conditions. N-type doping with e.g. polyethyleneimine (PEI) was shown in literature for single tube and network TFTs [68].

2.2. Transparent conductive films and their figures of merit

A transparent conductive film (TCF) is a thin-film which has a certain transmittance T in the optical range and is at the same time electrically conducting. The more general term transparent electronic conductor (TEC) can be used synonymously. In this work, T (in the range of 0 to 1) is the average value extracted from spectral measurements at a wavelength of (550 ± 5) nm. The sheet resistance $R_\square$ in units of $\Omega/\square$ is a common unit for the electrical DC resistance of such a thin layer. Evidently both values are depending on the thickness of the layer. A thinner layer will give a higher transmittance but also a higher sheet resistance. A thicker layer will better conduct the electrical current but absorb more light. It is therefore important to always look at both values in combination when comparing the electro-optical performance of different TCFs.

Two different figures of merit (FOM) that both combine T and $R_\square$ to a single quantitative value are commonly used when comparing transparent electronic conductors or TCFs. Already in 1976 Haacke introduced the figure of merit φ_{TC} for being able to more easily compare the performance of thin metal and oxide TECs which is defined as follows [39]:

$$\varphi_{TC} = T^{10}/R_\square \tag{2.28}$$

He chose the exponent of 10 to achieve a maximum FOM at about 90 % optical transmission. Another exponent could be chosen to adapt to a specific need.

The other figure of merit often used in the context of TCFs made of CNNs, graphene or Ag nanowires is the ratio of the electrical DC conductivity σ_{DC} and the optical conductivity σ_{op}. As derived in [100] the ratio can be correlated to T and $R_\square$ by the following formula:

$$T = \left(1 + \frac{188\,\Omega}{R_\square} \frac{\sigma_{op}}{\sigma_{DC}}\right)^{-2} \tag{2.29}$$

This is equivalent to the dimensionless expression:

$$\frac{\sigma_{DC}}{\sigma_{op}} = \frac{\sqrt{T}}{1 - \sqrt{T}} \frac{188\,\Omega}{R_\square} \tag{2.30}$$

Figure 2.17 shows a typical electro-optical curve of T vs. $R_\square$ (blue solid line) for materials used in this work. The FOM φ_{TC} calculated for the complete range is shown in fig. 2.17a, the values for σ_{DC}/σ_{op} are

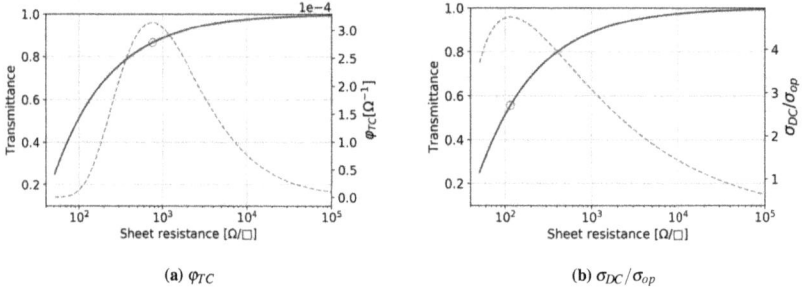

(a) φ_{TC} (b) σ_{DC}/σ_{op}

Figure 2.17.: Comparison of two different figures of merit for a typical curve of T vs. $R_\square$ (blue solid line on primary y-axis) for CNNs used in this work; the red dashed line is the calculated figure of merit (secondary y-axis), the red circle shows the combination of T vs. $R_\square$ at the maximum value for each FOM.

plotted in fig. 2.17b (dashed red lines). The red circle indicates the combination of T and $R_\square$ resulting in the highest values for both FOMs. Obviously σ_{DC}/σ_{op} favors a lower sheet resistance which leads to a relatively low transmittance for display applications. φ_{TC} gives maximum values for the realized TCFs that are at an optical transmission close to 90 % which confirms the validity of the chosen exponent discussed above. It therefore seems to be the more reasonable metric for the given application.

2.3. Thin film transistors

The realization of the semiconductor in a thin-film transistor (TFT) is one of the main applications of carbon nanotube networks (CNN) in this thesis. This section gives a short introduction into the technological realization and the device physics.

2.3.1. Thin film transistor setup

The thin-film transistor became popular with the invention of active matrix liquid crystal displays in the 1970s. It is a field effect transistor with the three terminals source, drain and gate similar to the well known metal-oxide-semiconductor field effect transistor (MOSFET). There are several possibilities how a TFT can be built. A general classification is made regarding the position of the gate contact. As a consequence we distinguish between top or bottom gate devices. In this work only bottom gate devices are realized. The following sections apply however to both categories.

A second classification is made depending on the position of semiconductor and source and drain contacts (see fig. 2.18). When both contacts and the semiconductor are in direct contact with the dielectric it is a coplanar or bottom contact device. The second variation has the source and drain contacts on top of the semiconductor and is therefore called staggered or top contact device.

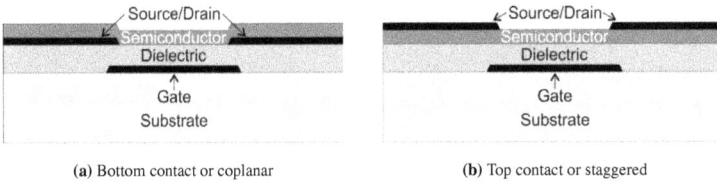

(a) Bottom contact or coplanar   (b) Top contact or staggered

Figure 2.18.: TFT setups with bottom gate configuration.

The schematic of a TFT with its three terminals and the usually applied voltages is shown in fig. 2.19.

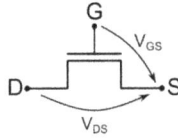

Figure 2.19.: Symbol of a TFT with terminals G = gate, D = drain and S = source plus corresponding applied voltages.

2.3.2. Thin film transistor characteristics

Although carbon nanotube network TFTs have different conduction mechanisms, they behave similar to MOSFETs. In first approximation the classical MOSFET equations are therefore used to describe their transfer and output characteristics. Three different regimes are distinguished.

I: Below the threshold voltage V_{th} the channel is cut off and there is no conduction. In reality there is a small leakage or off current I_{off}. This is due to the fact, that these transistors do not have p-n junctions and the channel cannot be fully depleted of charges (see eq. (2.31)).

II: For gate-source voltages V_{GS} much higher than the drain-source voltage V_{DS} the current through the channel increases linearly with V_{GS}. It is therefore called the linear region (see eq. (2.32)).

III: When the gate-source voltage is small compared to the drain-source voltage V_{DS} the channel is pinched off. The transistor is in saturation (eq. (2.33))

The gate-source voltage at which the conducting channel changes form depletion to conduction is called threshold voltage V_{th}. In consequence the three aforementioned regimes are for a fixed V_{DS} not only depending on V_{GS} but rather the difference $V_{GS} - V_{th}$. This becomes also visible in eqs. (2.31) to (2.33). CNT transistors behave as p-type devices under ambient conditions and without further modifications. The MOS-FET current equations are therefore given for the case of a p-type transistor, with negative voltages and currents:

$$ \text{I} \qquad I_{DS} = I_{\text{off}} \approx 0 \qquad\qquad V_{GS} - V_{th} \geq 0 \qquad \text{cut off} \qquad (2.31) $$

$$ \text{II} \qquad I_{DS} = -\beta \left[(V_{GS} - V_{th}) V_{DS} - \frac{V_{DS}^2}{2} \right] \qquad V_{GS} - V_{th} < V_{DS} < 0 \qquad \text{linear} \qquad (2.32) $$

$$ \text{III} \qquad I_{DS} = -\frac{\beta}{2} (V_{GS} - V_{th})^2 \qquad\qquad V_{DS} < V_{GS} - V_{th} < 0 \qquad \text{saturation} \qquad (2.33) $$

With I_{DS} = drain-source current, I_{off} = off-current, V_{GS} = gate-source voltage, V_{th} = threshold voltage and V_{DS} = drain-source voltage. The factor β is technology dependent:

$$ \beta = \mu_p C_G \frac{W}{L} \qquad (2.34) $$

with μ_p = charge carrier mobility for holes, W_C = channel width, L_C = channel length and the specific gate capacitance

$$ C_G = \frac{\varepsilon_0 \varepsilon_r}{d_{\text{diel}}} \qquad (2.35) $$

where ε_0 and ε_r are the vacuum and relative dielectric permittivity, respectively and d_{diel} is the thickness of the dielectric. The form of the resulting idealized transfer characteristics (I_{DS} over V_{GS}) and output charactersitics (I_{DS} over V_{DS} for several V_{GS}) is shown in fig. 2.20 as both linear and logarithmic plot.

2.3.3. Characterization of thin-film transistors

2.3.3.1. Electrical TFT measurements

The transfer and output characteristics of a single TFT are measured using a semiconductor analyzer equipped with three source-measure-units (SMU). Each SMU can source a DC voltage or current and measure at the same time voltage and/or current. With the help of a probe station and probes mounted on micro manipulators source, drain and gate are connected with one SMU each.

In the transfer characteristics measurement the drain is fixed on ground (0 V) and the source is put on a constant potential anywhere from -0.1 V to -10 V. A voltage sweep is then applied to the gate that might reach from -20 V to 20 V with steps from 0.1 V to 1 V. Ideally the curves for a gate sweep in positive direction should be identical to a sweep in negative direction. As this is not always the case both directions should be scanned.

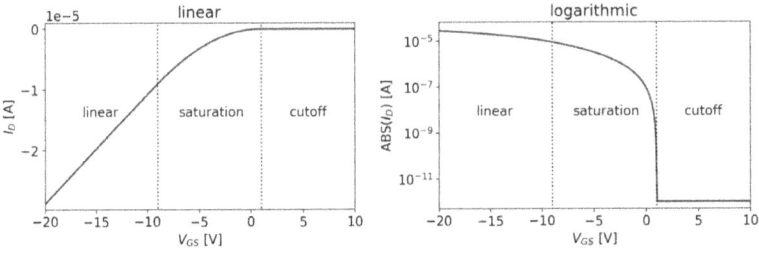

(a) transfer charactersitics; left: linear plot, right: logarithmic plot of abs(I_{DS})

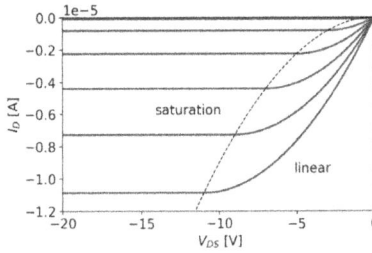

(b) output charactersitics for $V_{GS} = -10\,\text{V}$ to $2\,\text{V}$ in steps
of $2\,\text{V}$

Figure 2.20.: Idealized transfer and output characteristics of a p-type TFT with indicated conduction regimes for V_{th}=1 V, W_C=200 µm, L_C=10 µm, d=70 nm, μ_p=0.1 cm²/(V s) and $\varepsilon_r = 7.1$.

For output characteristic measurements the drain is again kept at ground level and a constant DC voltage is applied to the gate. The voltage sweep of e.g. 1 V to -10 V is then applied to the source. This can be repeated for a set of gate voltages outside the cutoff region to get an array of output curves for the complete operating range.

In particular, during technology development it is advised to measure the current at all three nodes during such measurement to distinguish between the current flowing in the actual channel and undesired leakage currents through the dielectric or outside the TFT channel area.

For the characterization of TFT performance some common figures of merit are used that can be extracted from transfer or output characteristics. In this work the following figures of merit are used:

2.3.3.2. On-current, off-current and on/off-ratio

The on-current I_{on} is the maximum drain-current that the TFT is able to deliver. Obviously this is depending on the scanned V_{GS} range and the applied V_{DS}. The off-current I_{off} on the other hand is the drain-current in the cutoff region. The on/off-ratio is defined by these two values by

$$I_{on}/I_{off} = \left| \frac{I_{on}}{I_{off}} \right| \tag{2.36}$$

and defines how efficient the drain current can be modulated by the gate-source voltage. From eqs. (2.32) and (2.33) it becomes obvious that the amount of I_{DS} and therefore I_{on}/I_{off} can be maximized by increasing β. This is achieved by optimizing the TFT technology to have

- a maximum W/L ratio while keeping a small TFT footprint which is especially important in the case of transmissive AMLCDs
- choosing a dielectric with high ε_r while staying compatible in material and process conditions to substrate and other applied layers
- decreasing the thickness of the dielectric d_{diel} while assuring a pinhole-free layer which supports the applied electric field strengths without significantly increasing leakage currents.

2.3.3.3. Threshold voltage

The threshold voltage V_{th} defines the gate-source voltage at which the TFT begins to accumulate enough charge carriers in the channel to render it conducting. There is a range of methods to extract V_{th} from measurement data [95]. The most common ones use the linear or saturation regime in the transfer characteristics.

Saturation regime The plot $\sqrt{I_D}$ vs V_{GS} ideally leads to a straight line in the saturation region (see fig. 2.21). A line of best fit along the straight part of the square root transfer curve (usually fit in the point of maximum slope) reveals the threshold voltage at the intersection with the x-axis. According to eq. (2.33) the threshold voltage is extracted from the fit line as

$$V_{th\text{-}sat} = V_{GS}\big|_{\sqrt{I_D}=0}. \tag{2.37}$$

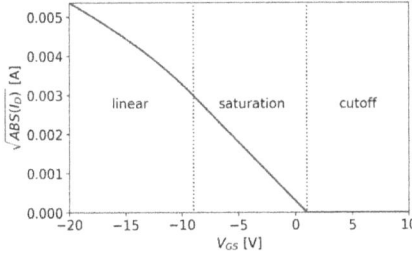

Figure 2.21.: Square root plot of abs(I_D) vs. V_{GS}, same data as in fig. 2.20.

Linear regime   A similar approach can be taken in the linear regime of the transfer curve. A straight line of best fit along the linear section of the I_D vs V_{GS} plot (compare fig. 2.20a) reveals the threshold voltage according to eq. (2.32) as

$$V_{\text{th-lin}} = V_{GS}|_{I_D=0} - \frac{V_{DS}}{2}. \tag{2.38}$$

2.3.3.4. Sub-threshold swing

Just below the threshold voltage the amount of drain current rises exponentially with the gate-source voltage shifting in direction of the on-state, resulting in a straight line in the logarithmic plot of transfer characteristics. The inverse slope of this straight is called sub-threshold swing S and is measured in V/decade. The sign is not really important as only the gradient counts. It can be extracted from transfer characteristics with a fitting straight

$$S = \left| \frac{dV_{GS}}{d \log I_D} \right| \tag{2.39}$$

and is an often used figure of merit for FET performance. The smaller S, the quicker the transistor current rises by a gate stimulus, which is a measure for quality. From theoretical MOSFET calculations it can also be expressed as:

$$S = \frac{k_B T}{q} \left(1 + \frac{C_{IT}}{C_{\text{diel}}} \right) \ln 10 \tag{2.40}$$

with k_B the Boltzmann constant, T temperature, q charge of an electron, C_{IT} capacitance caused by interface traps and C_{diel} the capacitance of the dielectric [16]. Even though the transport mechanisms are not necessarily identical, S is also applied to characterize CNN-TFTs [78]. From eq. (2.40) it can be easily seen why S serves as FOM. With increasing number of trap states at the interface between dielectric and semiconductor (which here implicitly also contains trap states in the dielectric or the CNN themselves), the subthreshold swing increases. The TFT performance can therefore be improved maximizing gate capacitance and minimizing the number of interface traps.

The theoretical limit of S for $C_{\text{diel}} >> C_{IT}$ is at about 60 mV/decade at room temperature which can be almost met by high quality MOSFETs. a-Si TFTs typically are in the range from 0.3 V/decade to 0.5 V/decade [146]. Single CNT transistors can achieve 60 mV/decade [62].

2.3.3.5. Charge carrier mobility extraction

The charge carrier mobility μ in the semiconducting TFT channel can be extracted with similar methods as described above for V_{th}. The same fitting straights are used.

Saturation region   We call the slope of the fitting straight in the square root plot m_{sat} as it fits the curve in the saturation region. Once it is determined the charge carrier mobility in saturation can be calculated using the TFT dimensions and dielectric constant of the dielectric as follows.

$$\mu_{sat} = \frac{L_C}{W_C} \frac{2m_{sat}^2 d_{diel}}{\varepsilon_0 \varepsilon_r} \tag{2.41}$$

Linear region   For determination of the mobility in the linear region we use the slope of the fitting straight m_{lin}. For a given V_{DS} the charge carrier mobility can be calculated with the following equation.

$$\mu_{lin} = \frac{L_C}{W_C} \frac{m_{lin} d_{diel}}{\varepsilon_0 \varepsilon_r V_{DS}} \tag{2.42}$$

The extraction from the linear region is mostly used in this work as V_{DS} is often kept low to avoid charging effects in the dielectric.

Both eqs. (2.41) and (2.42) are based on the model of a bulk semiconductor with a parallel plate capacitance for the area $W_C \cdot L_C$. In the case of a CNN this model is not necessarily correct as the channel area might be covered with a sparse network leading to individual current paths instead of a bulk semiconductor that fills the complete area. The calculated values do therefore not reflect the intrinsic mobility in the SWNTs but give an **effective device** charge carrier mobility μ_{device}. This further means that devices with identical dimensions might have different effective mobilities depending on the network density D. To extract intrinsic mobilities Yoon et al. used e.g. C-V and I-V measurements [144]. The effective device mobility is however a good indicator for comparing the TFT performance with other technologies that use the same device area and it is used throughout this work.

2.4. Liquid crystal displays

Liquid crystal displays (LCD) started the era of flat planel displays and are still the dominating species although recently organic light emitting diode (OLED) displays take more and more market shares. In the previous decades various types of liquid crystals and different driving methods were developed. During a long period the twisted nematic (TN) LCD was the commercially most important technology. It is nowadays more and more replaced with in-plane switching and vertically aligned LCD modes due to improved optical performance and/or higher switching speeds. The TN technology was nevertheless chosen for realizing the demonstrators of this thesis as it is simpler in realization. In any case the developed transparent conductive films can be applied to all of the mentioned technologies. Because of its even lower complexity also polymer dispersed liquid crystal displays (PDLCD) were used as first demonstration step. The general setup of these LCDs and different addressing methods to electrically drive the displays will be discussed in the following sections.

2.4.1. Nematic liquid crystals

In a solid crystal, molecules have a well defined order within the crystal lattice. Once this crystal is heated over its melting point it's molecules usually take a direct transition into an isotropic liquid phase. In this liquid phase the molecules follow only Brownian motion and there exists no long-range order as in the solid state before. The therm "liquid crystal" therefore seems to be contradictory. In 1888, Reinitzer discovered however a group of materials that exhibit one or several additional phases between the solid and the isotropic liquid state [17]. Until today various types of liquid crystal molecules which can have several "liquid crystal phases" were discovered. In this thesis only calamitic liquid crystal mixtures in the nematic phase are used. Its properties are quickly introduced in this chapter.

The group of calamitic liquid crystals consist of rod-like molecules as simplified depicted in fig. 2.22a. They have rotational symmetry along the long molecule axis. The orientation of a calamitic LC molecule is described by the $\vec{n}$-director, which points into the direction of the long molecule axis. Due to the polarizability of the molecule's core they exhibit an optical anisotropy resulting in different refractive indexes parallel ($n_\parallel$) and perpendicular ($n_\perp$) to the $\vec{n}$-director. This birefringence is given by eq. (2.43).

$$\Delta n = n_\parallel - n_\perp \tag{2.43}$$

In the presence of an electrical field the mentioned polarizability also leads to a dielectric anisotropy which is determined by eq. (2.44).

$$\Delta \varepsilon = \varepsilon_\parallel - \varepsilon_\perp \tag{2.44}$$

While $\Delta n > 0$ holds always for calamitic LCs, $\Delta \varepsilon$ can have negative and positive values depending on the molecule structure. In this thesis only LC mixtures with positive $\Delta \varepsilon$ are relevant.

From the many phases a liquid crystal can have between the solid and the liquid state, here only the nematic phase is relevant. In this phase the molecules have lost their long range positional order but they attempt to keep the same orientation. This means that the long axis of all molecules points roughly in the same direction resulting in an average phase n-director $\vec{n}_{phase}$ (see fig. 2.22b). Due to this orientational order the bulk nematic phase inherits the optical properties of the individual molecules and is birefringent.

The orientation of the molecules can be influenced by an electrical field. For LCs with $\Delta \varepsilon > 0$, the molecules attempt to align parallel to the electrical field lines. Since the polarization charges in the molecule's core have much higher mobility than the molecule itself, the orientation is independent of the polarity of the

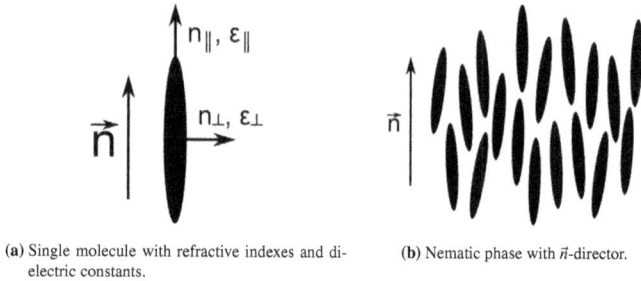

(a) Single molecule with refractive indexes and di-   (b) Nematic phase with $\vec{n}$-director.
electric constants.

Figure 2.22.: Simplified representation of calamitic liquid crystal molecules.

electrical field. Usually LCDs are addressed by AC signals in the kHz range. Longer time DC voltages can lead to decomposition of the LC molecules and ion migration. Both would deteriorate the display effect. Of course the electrically induced forces have to compete against existing mechanical forces like the one coming from the above mentioned orientational order combined with the viscosity of the mixture or anchoring forces between molecules and solid surfaces.

Commercial nematic LCs are always a mixture of different molecule types (up to 20 or more). The type and percentage of the ingredients is optimized to achieve a product that fulfills the needs for Δn, $\Delta \varepsilon$, working/storage temperature range, bulk resistivity, switching speed etc.

2.4.2. Polymer dispersed liquid crystal displays

Polymer dispersed liquid crystal displays have a relatively simple setup and were therefore chosen as a basic testing vehicle for transparent electrodes made of carbon nanotubes. Since they don't rely on polarizing filters they have a very high transmission. Their limited contrast ratio makes them however less favorable for high quality information displays. As can be seen in fig. 2.23 a PDLCD consists of two sandwiched transparent substrates with the transparent conductive film (TCF) at the inside of the resulting cell. In between are liquid crystal filled micro-spheres in a polymer matrix. In this work the used LC mixture is nematic as described in section 2.4.1.

The liquid crystal molecules in a micro-sphere are aligned and point in a certain direction which is indicated by the n-director $\vec{n}_{sphere}$. Without any applied electrical field the directors of different micro-spheres are randomly dispersed (see fig. 2.23a). Light passing through this cell is being scattered resulting in an milky white appearance [18, chapter 7.3.7]. When applying an electrical field to the electrodes with sufficient amplitude the liquid crystal molecules align parallel to the field lines (fig. 2.23b). If the refractive index of the polymer n_p is well matched to $n_{\|}$ of the LC mixture, such a PDLCD can have an almost 100 % transmittance.

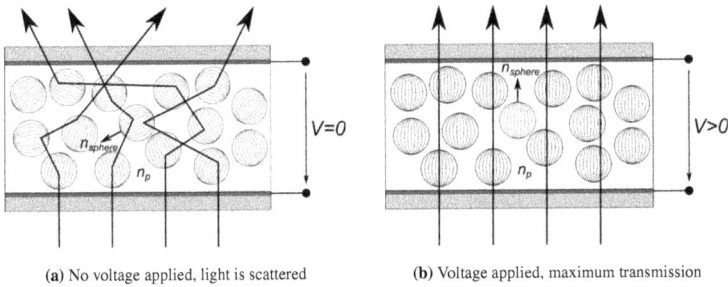

(a) No voltage applied, light is scattered   (b) Voltage applied, maximum transmission

Figure 2.23.: Working principle of a PDLCD consisting of LC micro-spheres in polymer matrix, sandwiched between substrates with TCF; n_p refractive index of polymer matrix.

2.4.3. The twisted nematic liquid crystal display

The twisted nematic LCD was invented in 1970 by Wolfgang Helfrich and Martin Schadt [47, 114] and was up until recently the most used type of LC displays. The general setup is displayed in fig. 2.24. Besides substrates and transparent conductive film as in the PDLCD, the sandwich of a TN cell also contains a thin orientation layer on top of both TCFs and polarizing filters on the outside.

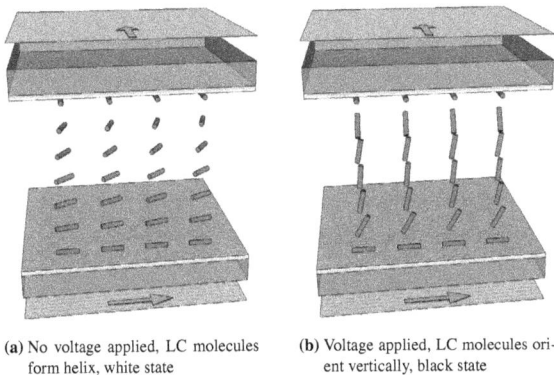

(a) No voltage applied, LC molecules   (b) Voltage applied, LC molecules ori-
form helix, white state   ent vertically, black state

Figure 2.24.: Working principle of a 90° twisted nematic LCD with crossed polarizers; shown are the LC molecules represented by rods, sandwiched in between substrates with TCF and orientation layer; flashes indicate orientation of rubbing direction and outside polarizing filters.

The orientation layer is commonly realized as thin polyimide layer that is rubbed with a velvet cloth. The rod-like LC molecules close to the surface align parallel to this rubbing direction. When the rubbing direction of the two orientation layers on back- and frontplane do not point into the same direction, the LC molecules form a helix due to their orientational order. In a standard TN cell this twist is 90° (see fig. 2.24a).

A dopant in the LC mixture assures the rotation direction in which the helix forms. Further the molecules next to the orientation layer have a small tilt angle in the range of 1° to 2°. This means that one end of the molecule is closer to the surface than the other. This pre-tilt is mostly influenced by the type of PI and rubbing technique and is important for a homogeneous switching behavior of the LCD.

The birefringence of the LC layer in such a TN cell alone does not give any visible effect. It only affects the polarization of the passing light. Therefore polarizing filters are added so that the complete stack can be used as a light valve. The first filter acts as polarizer, while the second one acts as analyzer.

The transmission through a TN cell in the off-state (no voltage applied) with 90° twist and perfect parallel aligned polarizing filters can be described with the Gooch-Tarry equation [37]:

$$T_{TN} = \frac{\sin^2(\frac{\pi}{2}\sqrt{1+u^2})}{1+u^2} \tag{2.45}$$

with

$$u = 2 \cdot \frac{\Delta n \cdot d}{\lambda} \tag{2.46}$$

where d is the thickness of the LC layer and λ is the optical wavelength. The plot of eq. (2.45) in function of u is shown in fig. 2.25. The so-called Gooch-Tarry minima in this plot can be parametrized with the following equation.

$$(\frac{\Delta n \cdot d}{\lambda})^2 = m^2 - \frac{1}{4}, m \in \mathbb{Z} \tag{2.47}$$

A maximized contrast ratio is achieved when the display fulfills this criterion and works in one of these minima in its non-addressed condition. The optimum thickness for such a TN cell can be derived thereof for a given Δn and a choice of m. It needs to be noted however that this holds in principle only for a certain wavelength. Usually the optimization is done for an average wavelength of $\lambda = 550$ nm. For improved switching speed and low addressing voltages commercial TN displays are usually optimized for the first or second minimum ($m = 1, 2$).

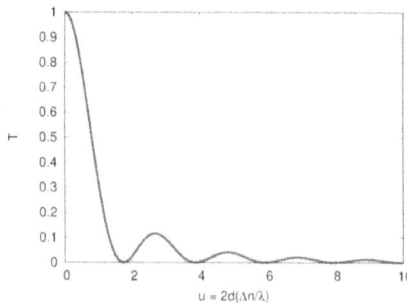

Figure 2.25.: Gooch-Tarry graph of a 90° TN cell in normally black configuration.

The physical effect behind this formula is the change of polarization that the passing light experiences. At the first polarizing filter only light with polarization parallel to the filter can pass. This results in 50% light loss for non-polarized light and is a big disadvantage of all display types that depend on polarizing filters. The photons first only experience the LC molecule's $n_\parallel$ as refractive index of the medium. When going deeper into the LC layer the orientation of the LC molecules changes with the helix. Therefore the photons start also to experience $n_\perp$. Due to Δn the polarization of the photons traveling through the LC layer changes with the way traveled. From linear polarization parallel to the filter it changes to elliptical polarization and ends for a well defined TN cell with linear polarization that is turned by 90° in reference to the entry of the cell. When the analyzing filter at the back of the cell is parallel to the front filter the light cannot pass, leading to a black pixel. This is the so-called normally black mode, described by eq. (2.45). For crossed polarizing filters the transmitted light can pass, resulting in a normally white pixel.

When applying an electrical field between the two electrodes the LC molecules experience a force that tries to reorient them parallel to the electrical field. Once a certain threshold voltage V_{th} is surpassed, the molecules in the volume start rotating leading to a diminution of the above described effect. Once the electrical field is high enough, all molecules besides the ones next to the substrate surfaces are parallel to the electrical field (see fig. 2.24b). The linearly polarized light entering the cell experiences no major change in the refractive index n and keeps its orientation. The normally-white cell becomes black, the normally-black cell becomes white. Figure 2.26a shows the change of transmission of a normally-white display in function of the applied addressing voltage including V_{th} that needs to be surpassed before seeing any change. Electrically speaking such an LC pixel is a parallel plate capacitor with the two TCFs as electrodes and the

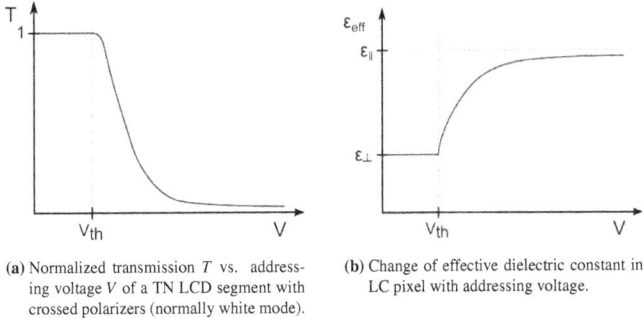

(a) Normalized transmission T vs. addressing voltage V of a TN LCD segment with
crossed polarizers (normally white mode).

(b) Change of effective dielectric constant in
LC pixel with addressing voltage.

Figure 2.26.: Voltage-dependent behavior of a TN-cell.

liquid crystal layer as dielectric. It's capacitance can therefore be calculated by the equation

$$C_{LC} = \varepsilon_0 \varepsilon_{LC} \frac{A}{d} \tag{2.48}$$

where ε_0 is the permittivity of the vacuum, ε_{LC} the dielectric constant of the liquid crystal layer, A the lateral area of the superimposed pixel electrodes and d the thickness of the LC layer or the cell spacing.

Under static conditions the display effect is purely voltage driven with basically no current flow besides negligible leakage currents caused by ionic impurities in the liquid crystal mixture. During pixel switching the charging and de-charging of the pixel requires of course some current flow.

As described above, the transmission is changed by reorientation of the LC molecules in combination with their Δn. Since the molecules also exhibit a $\Delta \varepsilon$, the effective dielectric constant of the pixel changes as well. In the off-state the molecules lie parallel to the electrodes resulting in $\varepsilon_{\text{eff}} = \varepsilon_\perp$. In the on-state most of the molecules are perpendicular to the electrodes with the effect that $\varepsilon_{\text{eff}} = \varepsilon_\parallel$. For the gray levels in between there is a gradual change as depicted in fig. 2.26b [18, section 7.1.4]. The pixel capacitance is therefore a function of the addressing voltage and eq. (2.48) can be re-written to eq. (2.49).

$$C_{LC} = \varepsilon_0 \cdot \varepsilon_{LC,\text{eff}}(V) \cdot \frac{A}{d} \qquad (2.49)$$

The first TN-LCDs were simple segmented displays used in pocket calculators and watches. The low power consumption and thin construction allowed completely new applications especially for portable devices. The direct addressing of segmented displays however limits scalability and increase in information density. Each element plus the counter electrode needs to be connected by a wire and individually addressed by a driver (see fig. 2.27a). By segmenting also the counter electrode and using multiplexing methods the number of external contacts can be reduced but the wires to each element stay a limiting factor.

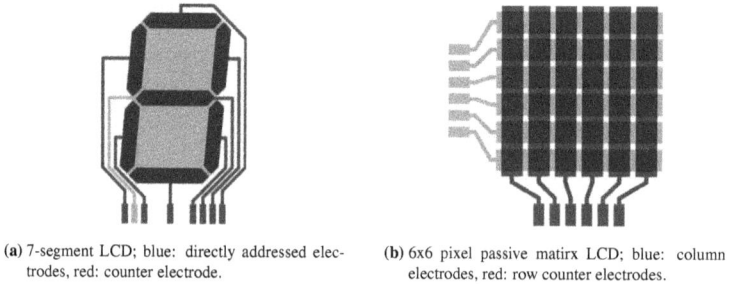

(a) 7-segment LCD; blue: directly addressed elec-   (b) 6x6 pixel passive matirx LCD; blue: column
trodes, red: counter electrode.   electrodes, red: row counter electrodes.

Figure 2.27.: Schematic layout of passively addressed LCDs.

2.4.4. The active matrix liquid crystal display

To improve the information density and versatility of liquid crystal displays, dot matrix addressing schemes were developed. So-called passive matrices are basically an array of pixels formed by crossed TCF bars on both front- and backplane, each crossing resulting in a square pixel (see fig. 2.27b).

A passive matrix with n lines and m columns results in $n \times m$ pixels with only $n + m$ contacts to a driver. Table 2.2 shows the multiplexing factor

$$\text{mux} = \frac{\text{pixels}}{\text{connections}} \qquad (2.50)$$

for different display resolutions.

Table 2.2.: Pixels, connections and multiplexing factor for a display matrix with m columns and n lines. In the case of color displays the columns are tripled to have RGB subpixels.

m	n	pixels	connections	mux
16	16	256	32	8
64	64	4096	128	32
256	256	65536	512	128

A passive matrix LCD is fairly simple to produce. The fact, that all pixels are electrically connected via the lines and columns demands however sophisticated addressing schemes and cross-talk limits resolutions to below the VGA resolution of 640 x 480 pixels. In consequence active matrix liquid crystal displays (AMLCD) were developed where each individual pixel is controlled by a transistor. This allows the realization of large area screens with high definition resolutions and above. It complicates the production process however significantly by introducing several additional layers which are usually deposited by vacuum processes like sputtering or plasma-enhanced chemical vapor deposition (PECVD). The deposited layers further need to be patterned by photolithography and wet or dry etching processes (see section 6.5.3.1).

The equivalent circuit diagram of an AMLCD is shown in fig. 2.28. The main optical element is the pixel which is represented by the pixel capacitance C_{LC}. It is realized by the liquid crystal layer sandwiched between two transparent electrodes. The backside electrode is patterned and connected to the active matrix on a pixel level, while the frontside electrode is the unpatterned counter electrode CE that is common for all pixels. The TFT connects the backside pixel electrode to the column busline which is also called source line. The TFT acts basically as a switch that is remote-controlled by the row busline, also called gate line.

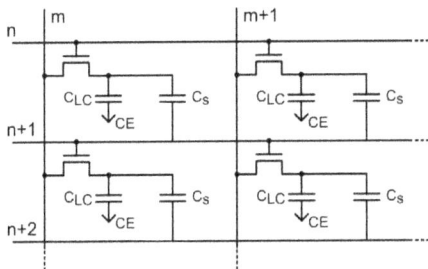

Figure 2.28.: Structure of an AMLCD with pixel capacitance C_{LC}, connected to the counter electrode CE and storage capacitor C_s.

The image is written into the display by using the "line at a time" method. A complete line of pixels is activated by rendering the TFTs conductive. This is done by applying an appropriate "on-potential" V_{gON} to the gate line. At the same time all the TFTs of the other rows are rendered nonconducting by applying the "off-potential" V_{gOFF}. The pixel capacitance is then charged to an analog gray level voltage applied to the source line. This is done in parallel for all pixels of the activated row. Once the pixels are charged to the appropriate voltage the row is deactivated, the next one is activated and a new set of gray levels can be sent to the next pixel row. When the last gate line is addressed the next frame is written by beginning from the first line.

To achieve a DC-free LC driving the pixel voltages are inverted in two consecutive frames. In larger screens this is usually done by keeping the counter electrode on a fixed intermediate level and having the pixel voltage swing symmetrically around this fixed voltage. For example if a fully black pixel would require a pixel voltage of $V_{LC} = 5\,\mathrm{V}$ the counter electrode could be fixed at $V_{CE} = 6\,\mathrm{V}$ while having V_{LC} change between 1 V and 11 V.

Parasitic capacitances in the matrix lead to voltage shifts in the pixel, most remarkably the kickback voltage caused by the TFT's gate-drain capacity C_{GD} [18, section 7.4.1.5]. These voltage shifts can to some degree be compensated by adjusting V_{CE} and an asymmetric gamma curve for the positive and negative branch. The variable pixel capacitance makes the level shifts however gray-level dependent. An additional fixed storage capacitor C_s (see fig. 2.28) can reduce the variability of the C_{LC}. It is realized with the two metal layers of the TFT and it's gate dielectric and is connected parallel to C_{LC} with the other electrode connected to the following gate line. C_s usually has a capacity in the same order of magnitude as the pixel itself. Another advantage of the increased effective pixel capacity is a decreased effect of leakage currents on the pixel voltage.

2.4.5. Demands on a thin-film transistor in an active matrix

In a transmissive AMLCD, the pixel size defined by the busline pitches of rows and columns, needs to be shared by the TFT, buslines, LC-pixel, storage capacitor and spacing between these elements to avoid short circuits and limit cross-talk (compare fig. 6.16 in section 6.5.3.1). All parts besides the LC-pixel electrode are already opaque or even need to be shielded to avoid uncontrolled light leakage. The ratio between optically active part and the total pixel size gives the aperture ratio AR.

$$\mathrm{AR} = \frac{\text{optically active pixel area}}{\text{total pixel area}} \qquad (2.51)$$

To achieve a somewhat reasonable transmission and therefore contrast ratio of the display, all opaque parts need to be kept as small as possible. The footprint of the TFT therefore is a crucial factor.

As described in the above section the transistors in an active matrix are used as switches between source line and individual pixels. Ideally the channel would have zero resistance in the on-state and zero conductivity in the off state. Thin film transistors with a minimized footprint are however far from these specifications. The following simplified calculation will show what criteria in terms of on- and off-current, a transistor in an AMLCD needs to fulfill. For an optimized product, simulations are mandatory that include dynamic behavior of the complete active matrix.

The AMLCD is continuously refreshed with a frame rate f_{frame}. The time that is available to address one line is therefore

$$t_{\text{line}} = \frac{1}{f_{\text{frame}} \cdot N} \tag{2.52}$$

where N is the number of lines. When the specific line is activated the TFT is rendered conductive and the pixel is charged to the desired voltage via the source line and the TFT channel. This charging needs to be achieved during t_{line} and therefore demands a minimum on-current. For the rest of the frame time, which is

$$t_{\text{hold}} = \frac{N-1}{f_{\text{frame}} \cdot N} \tag{2.53}$$

the voltage level or more correctly the pixel charge needs to be stored. Since a non-ideal TFT has a certain off-current the charge in the pixel decreases during t_{hold}. This charge loss results in a voltage level drop. The acceptable drop is demanding a maximum off-current of the TFT and is depending on the steepness of the electro-optical behavior of the LC (compare fig. 2.26a) and the thereof resulting voltage step for one gray level.

The equivalent circuit for both charging and charge conservation is displayed in fig. 2.29 and reflects a simple RC-circuit where the TFT's channel is replaced by a resistor with two different values for on and off state. Any leakage currents caused by a non-ideal LC pixel, indicated in fig. 2.29 by a resistor parallel to C_{LC}, are neglected in the following calculations. The time constant of the RC circuit is given by eq. (2.54).

$$\tau_x = R_x \cdot C_{LC}, \qquad x \in \{\text{charge, hold}\} \tag{2.54}$$

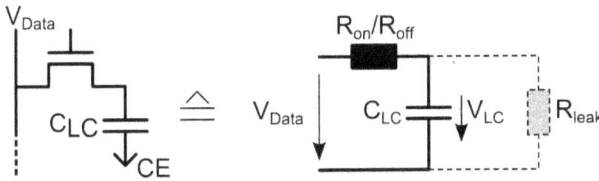

Figure 2.29.: Equivalent circuit of the LC pixel during charging and decharging via the TFT.

In first approximation C_{LC} is fully charged after 5τ. For pixel charging the minimum TFT resistance can therefore be calculated.

$$t_{line} > 5\tau = 5R_{\text{charge}} \cdot C_{LC} \tag{2.55}$$

$$R_{\text{charge}} < \frac{t_{line}}{5 \cdot C_{LC}} \tag{2.56}$$

The minimum on-current that the TFT needs to be able to deliver can therefore be specified as

$$I_{ON} > \frac{\Delta V_{LCmax}}{R_{charge}} \tag{2.57}$$

$$> \frac{5 \cdot \Delta V_{LCmax} \cdot C_{LC}}{t_{line}} \tag{2.58}$$

where ΔV_{LCmax} is the maximum voltage difference that will be applied to C_{LC}. During the hold time C_{LC} changes its charge via R_{hold} in reference to a series of different source voltages V_s that are applied to the column during addressing of pixels in other rows. A certain degree of leakage cannot be avoided. A crucial parameter is therefore the voltage drop in the pixel during the hold time.

$$\Delta V_{LC,hold} = \left| V_{LC(t=0)} - V_{LC(t_{hold})} \right| \tag{2.59}$$

The maximum off-current can therefore be defined as

$$I_{OFF} < \frac{\Delta V_{LC,hold} \cdot C_{LC}}{t_{hold}}. \tag{2.60}$$

Finally the necessary on/off-ratio can be derived from eqs. (2.58) and (2.60):

$$\frac{I_{ON}}{I_{OFF}} = \frac{5 \cdot \Delta V_{LCmax} \cdot C_{LC} \cdot t_{hold}}{t_{line} \cdot \Delta V_{LC,hold} \cdot C_{LC}} \tag{2.61}$$

which, by inserting $t_{hold} = (N-1)t_{line}$ can be rewritten to:

$$\boxed{\frac{I_{ON}}{I_{OFF}} = 5(N-1)\frac{\Delta V_{LCmax}}{\Delta V_{LC,hold}}} \tag{2.62}$$

The result shows that the necessary on/off-ratio linearly increases with the number of lines and is further only depending on the accepted $\Delta V_{LC,hold}$ and LC voltage range ΔV_{LCmax}.

For the AMLCD that will be presented in section 6.5.3 the liquid crystal saturates at about 5 V. With the positive and negative addressing cycles this leads to $\Delta V_{LCmax} = 10\,V$. The pixel capacitance is $C_{LC} + C_s = 0.6\,pF$ and the number of lines $N = 240$. With a frame rate of 50 Hz the currents can be calculated to $I_{ON} = 360\,nA$, $I_{OFF} = 3\,pA$ and hence $\frac{I_{ON}}{I_{OFF}} = 1.2 \times 10^5$.

2.5. Characterization methods

In the frame of this work a variety of characterization methods were used. Most of them are well known and the interested reader is encouraged to find further details in literature. This holds especially for the microscopy methods like optical microscopy, scanning electron microscopy (SEM) and atomic force microscopy (AFM) [11].

2.5.1. Layer adhesion

The adhesion of thin films like CNNs or metals on the substrate surface or on other thin films can be tested with a so-called scotch tape test. A piece of scotch tape is laminated onto the layer to be tested and pulled off suddenly towards the surface normal. If parts of the layer or the complete film is removed with the tape, the adhesion force is probably insufficient. An unharmed layer is evidence of good adhesion. Evidently this simplified and purely manual scotch tape test is only capable to deliver qualitative conclusions. There are more sophisticated methods allowing for standardized and also quantitative characterization that were however not used in this work.

2.5.2. Sheet resistance and contact resistance

For the electrical characterization of CNNs the sheet resistance of the deposited layer is an important property. Two different methods are used to determine the sheet resistance. The first one uses a four point probe with equally spaced metallic probes with distance a (see fig. 2.30a). A constant current is forced between the outermost probes. The resulting potential difference can then be sensed between the innermost probes. As basically no current flows in the sensing circuit there is no voltage drop due to leads and contact resistance. For a laterally infinite thin layer the voltage drop can be calculated as [79]

$$V = \frac{I}{\pi} \ln 2 \cdot R_s. \tag{2.63}$$

In the measurement V and I are known and the sheet resistance can therefore be calculated as

$$R_s = \frac{V}{I} \frac{\pi}{\ln 2}. \tag{2.64}$$

For laterally finite layers with width w and length l eq. (2.64) stays applicable as long as $a \ll w,l$. If a approaches the layer dimensions the factor $\frac{\pi}{\ln 2}$ has to be adapted by numerical calculations or determined from charts [9].

The four point probe measurement allows for fast characterization of unpatterned, deposited layers and automatically eliminates the influence of contact resistance. The second method on the other hand is a useful test pattern in applications where the TCF is patterned and electrically contacted by metal buslines and allows to measure sheet and contact resistance at the same time. The test pattern consists of a stripe of TCF with defined width. Below or above are thin metal fingers in well defined distances (see fig. 2.30b). Consecutive resistance measurements between the first and following fingers can be plotted as resistance vs. distance. The resistance values include both the sheet resistance and the contact resistance. Due to variation of the finger distance the two can however be separated. The plot ideally shows a straight line. The slope of the line is due to the TCF's sheet resistance. When correlated to the TCF stripe width $R_\square$ can be directly

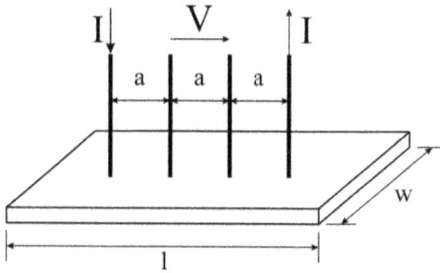

(a) Four point probe measurement with current forcing on outermost probes and voltage sensing on innermost probes.

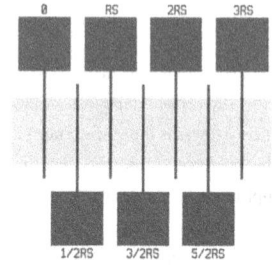

(b) Test pattern for sheet resistance measurement of the TCF strip with metal bars; the same structure can be used for contact resistance measurement between the specific metal and the layer.

Figure 2.30.: TCF characterization methods for sheet resistance and contact resistance to metal.

determined. The double contact resistance $2R_c$ can be extrapolated for finger distance $d = 0$. The specific contact resistivity ρ_c in $\Omega \cdot cm^2$ between TCF and metal can be determined by

$$\rho_c = R_c \cdot A \tag{2.65}$$

where A is the contact area between TCF and one metal finger.

3. Simulation of percolation in carbon nanotube networks using the Monte-Carlo method

The main building block in this work are networks of single-walled carbon nanotubes. The topography of these quasi-2D networks plays a crucial role for the different applications discussed in the following chapters. With SWNT diameters between 1 nm and 2 nm, the characterization of the CNN topography by classical optical microscopy is not possible. Even with more advanced techniques like scanning electron microscopy or atomic force microscopy it is hard to visualize individual SWNTs. Specifically, because the deposition is done on far from perfect substrates and surfaces (see section 7.2). For higher densities the individual nanotubes form an interwoven network of ropes where no beginning or end of tubes is visible. As a result of these difficulties a detailed characterization of CNN topography after deposition is not part of this thesis.

To get a better understanding on how these networks form and how metallic nanotubes influence the overall behavior, Monte-Carlo simulations of the network percolation were conducted as summarized in this chapter.

3.1. Setup of the Monte-Carlo simulation

The developed code simulates the formation of a CNN with its real-life dimensions where the individual tubes are realized as straight sticks. An area is defined, representing for example the channel of a TFT. The main outcome of the simulation is the percolation probability from one edge of the defined area to the opposite edge for mixed and metallic nanotubes depending on different starting conditions. For ease of analysis the network is represented in a rasterized manner. A flow chart of the main part of the simulation program is shown in fig. 3.1.

First the initial values are defined.

- dimensions of the channel area in width W_C and length L_C in μm
- length L_t of individual CNTs in μm

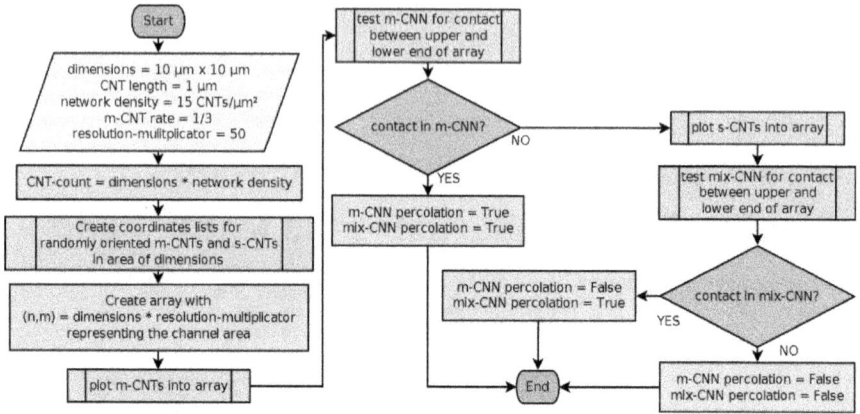

Figure 3.1.: Program flow chart of the main part; m-: metallic, s-: semiconducting, mix-: mix of m- and s-CNTs, CNN: carbon nanotube network; starting conditions are only for example.

- network density D in tubes/μm^2
- metallic content mc in percent (0 to 1)
- resolution-multiplicator M in pixel/μm

Based on these preconditions the total number of CNTs N_t in the channel area is calculated and a list of start- and end-coordinates for randomly oriented sticks for both m- and s-CNTs is created. First the m-CNTs are plotted into the (m,n) array. The plotting module based on the well known Bresenham algorithm [14] was adapted to have the following features.

1. **periodic boundary conditions** - Sticks leaving the array at one end re-enter on the other side. This allows for homogeneous CNT distributions with random start coordinates in the array
2. **semi-anti-aliasing** - The lines drawn by the Bresenham algorithm have a width of one pixel; at pixel steps this leads to non-touching segments (see blue pixels in fig. 3.2) that the following analysis algorithm would not count as connected elements. At these positions an additional pixel is added to keep the lines connected without increasing the line width more than necessary (yellow pixels in fig. 3.2)

The pixel clusters in the array are then detected and counted. Once a cluster is found that reaches from the bottom to the top edge of the array the flag for both the purely metallic network (m-CNN) as well as the mixed network of m-CNTs and s-CNTs (mix-CNN) are returned as TRUE. If the m-CNN does not show percolation the s-CNTs are added to the array with the still existing m-CNTs and the analysis is run once more to see if the mixed network connects the two edges or not.

The so far described sequence returns the percolation of m-CNN and mix-CNN for one specific network. To receive the percolation probability for a given parameter set the Monte-Carlo method [46] is used where a

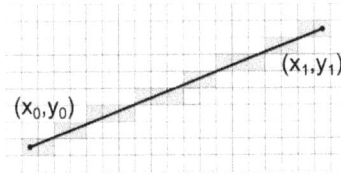

Figure 3.2.: Line plotting by Bresenham algorithm (blue pixels) and semi-anti-aliasing by connecting segments with additional (yellow) pixels.

large number of experiments (e.g. $N = 1000$ runs) with randomized start conditions (in this case the position and orientation of CNTs) are run. The percolation probability for m-CNN p_m and mix-CNN p_{mix} of a chosen density D are then defined as follows.

$$p_m = \frac{\sum \text{m-CNN contact}}{N} \tag{3.1}$$

$$p_{mix} = \frac{\sum \text{mix-CNN contact}}{N} \tag{3.2}$$

A more complete picture is received by scanning over a range of network densities.

The above described Monte-Carlo simulation is realized in Python 3. Although Python as interpreted programming language is not predestined for running computationally intensive Monte-Carlo simulations the use of efficient modules realized in C from the scientific computing libraries NumPy and SciPy as well as realizing the Bresenham algorithm as C-code with the module Cython drastically increases the speed compared to pure Python code. Nevertheless for large channel dimensions and/or high numbers of CNTs scanned over a wide range of network densities, the run time can go up to hours and repetitions for a variation of parameters might run for a day or more.

3.1.1. Graphical analysis of simulation runs

The percolation probabilities p_{mix} and p_m for mix-CNN and m-CNN vs. network density D are plotted in the upper part of fig. 3.3. All simulation parameters are given in the figure caption. From these curves one can extract D for a given percolation probability which is done by inverse search in combination with linear interpolation. A more compact representation of the simulation data is achieved by showing only D for p_{50} which represents 50% percolation probability plus the width of the transition zone in which $0 < p < 1$ by range bars. In order to have well defined limits $p_{0.5}$ and $p_{99.5}$ representing 0.5% and 99.5% percolation probability are chosen as upper and lower limit of these transition zones. This representation is shown in the lower part of fig. 3.3. The more important part when trying to realize a semiconducting device with mixed SWNTs is the gap between these two transition zones - the effectively semiconducting range. In this range of D the mix-CNN is always percolating while the metallic content, the m-CNN is never percolating. For

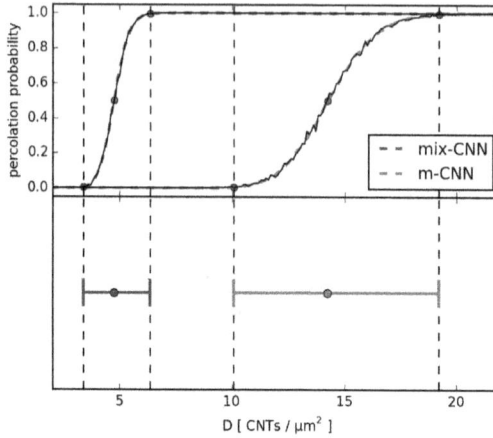

Figure 3.3.: Percolation probability plot for mix-CNN and m-CNN with extracted values for $p_{0.5}$, p_{50} and $p_{99.5}$ for both types of networks; $W_C = L_C = 10$, $L_t = 1$, $mc = 1/3$, $N = 8k$, $M = 50$; thin black lines for $N = 1k$.

easier description the transition zone of m-CNN TZ_m and mix-CNN TZ_{mix} plus the semiconducting range SR are defined as

$$TZ_m = p_{99.5m} - p_{0.5m} \tag{3.3}$$

$$TZ_{mix} = p_{99.5mix} - p_{0.5mix} \tag{3.4}$$

$$SR = p_{0.5m} - p_{99.5mix}. \tag{3.5}$$

While the data plotted in fig. 3.3 for $N = 8k$ shows smooth progress the overlayed thin black line for $N = 1k$ is significantly noisier. The overall form stays the same but local peaks might influence the detection of $p_{0.5}$ and $p_{99.5}$. Nevertheless $N = 1k$ is kept as standard parameter and smoothing of the data is achieved by using an algorithm published by Savitzky and Golay [113]. The Savitzky-Golay algorithm is a low pass filter. It subsequently calculates a smoothed value for each data point by fitting adjacent data points in a restricted window to a curve of chosen polynomial degree using linear least squares. Given the shape of the percolation probability curve a polynome of 2nd degree was chosen. The filter is implemented in the SciPy package. An example of smoothed data is shown in fig. 3.4.

3.1.2. Influence of pixel resolution

Given that the CNNs in this simulation are represented in a rasterized manner the pixel resolution has a significant influence on the results. As commercially available SWNTs have diameters between 1 nm and

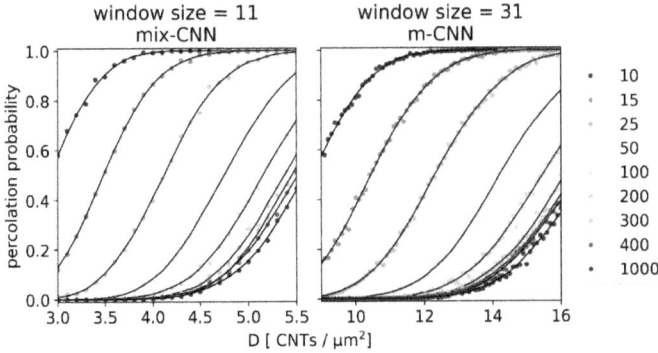

Figure 3.4.: Smoothing (black lines) of percolation probability data (points) with Savitzky-Golay algorithm, data points are fitted in moving window to 2nd order polynomial with indicated window size; $W_C = L_C = 10$, $L_t = 1$, $mc = 1/3$, $N = 1k$, M as parameter, data point distance on x-axis = 0.1.

2 nm the resolution of arrays representing the CNN should for a realistic representation have a resolution of at least

$$M > \frac{1\,\text{pixel}}{1\,\text{nm}} = 1000\frac{\text{pixel}}{\mu\text{m}}. \tag{3.6}$$

This results however in enormous matrix dimensions. Even a small channel area of $10\,\mu\text{m} \times 10\,\mu\text{m}$ would need to be represented by a matrix of $10'000 \times 10'000 = 10^8$ entries, consuming large amounts of memory and slowing down the cluster analysis of the created networks. Given the fact that a single simulation run results in hundred thousands of network analyses the overall runtime increases drastically with the matrix size.

On the other side too small a resolution overestimates the size of each SWNT leading to more connections between adjacent nanotubes. A comparison of identical simulation parameters but a resolution multiplicator M varied between 10 and 1000 is shown in fig. 3.5. It is actually the compacted representation of fig. 3.4. The extracted characteristic values are listed in table C.1 in appendix C.

Figure 3.5a shows the behavior as described above. For smaller M the percolation threshold is reduced for both m-CNN and mix-CNN. In all cases it holds however that $p_{50m} = 3 \cdot p_{50mix}$, which is not surprising given that there are always one third m-CNTs. The ratio of the transition zones $\frac{TZ_m}{TZ_{mix}}$ also stays close to 3. The semiconducting range SR is increasing for larger M and saturating towards higher values of M.

The square root of the runtimes for all simulation runs is shown in fig. 3.5a. The linear fit shows clearly that $t_M \propto M^2$ and can be explained by the fact that the number of matrix elements increases with the square of the multiplication factor. An increase in lateral precision is therefore bought by a quadratic augmentation of runtime.

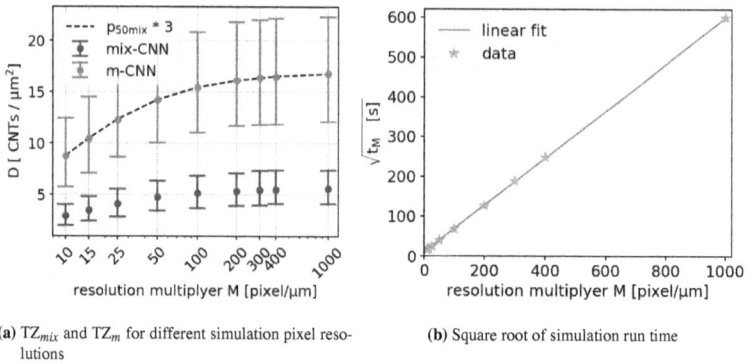

(a) TZ_{mix} and TZ_m for different simulation pixel resolutions

(b) Square root of simulation run time

Figure 3.5.: Simulation results for different resolutions multiplicators M;
$W_C = L_C = 10$, $L_t = 1$, $mc = 1/3$, $N = 1k$, $D \in [1, 25)$ in intervals of 0.1.

As a reasonable compromise a multiplication factor of $M = 50$ is generally used for all following simulations. This means that the values for percolation density found by simulation are generally lower than expected in reality. For the presented data there is an increase of 18% for p_{50} between $M = 50$ and $M = 1000$. Despite this discrepancy the overall behavior and trends when changing certain parameters should however stay comparable to reality.

3.2. Results of CNN percolation simulations

With the simulation tools developed in this work as presented in the beginning of this chapter, the influence of different parameters on the percolation behavior of CNNs will be described hereafter. If not mentioned otherwise, the following simulation parameters are used for the examples described below.

- pixel multiplication factor $M = 50$
- number of Monte-Carlo runs per network density $N = 1000$
- SWNT lenght $L_t = 1\,\mu m$
- metallic content $mc = 1/3$

3.2.1. Influence of channel geometry

The simulation results discussed so far are based on a square channel area of $W_C = L_C = 10\,\mu m$ which will further serve as reference. In an indefinite CNN of randomly oriented SWNTs with defined length the network density defines whether percolation takes place or not. For the given applications, the area is however confined and especially in the case of CNNs as semiconductors in a TFT channel, the miniaturization of the channel area is a crucial factor for the integration into display technology. Instead of general percolation behavior, the more important information is whether a (semi-)conducting network occurs between the upper and lower ends of the channel area. Changing the size and aspect ratio of this channel has therefore an influence on how the two sides are connected by the CNN.

For the first example, the square form of the channel is maintained, while the edge length is varied between 5 µm and 200 µm. The results are shown in fig. 3.6a. As the horizontal dashed lines clearly indicate, the 50% percolation probability of both mix-CNN and m-CNN do not change. The bigger the channel area however, the steeper are the percolation curves in the transition zones resulting in diminishing transition zones for bigger channel geometries. The bigger area therefore increases the averaging in the channel area, leading to a better defined percolation density. The result is an increasing effectively semiconducting range SR.

In the second example $W_C = 10$ µm was kept constant while L_C is varied between 5 µm and 200 µm (see fig. 3.6b). In this case p_{50mix} and p_{50met} are both increasing for longer channel lengths. The SR is increasing as well. Although the transition zones are not decreasing as strong as in the first example, SR for the different channel lengths is almost identical between examples 1 and 2.

In the third example the aspect ratios are reversed. $L_C = 10$ µm and W_C is varied between 5 µm and 200 µm (see fig. 3.6c). In this case p_{50mix} and p_{50met} are both decreasing for larger channel widths. The semiconducting range is also slightly increasing with W_C.

Interestingly enough, the aspect ratio of the channel area has also an influence on the form of the percolation probability curve, represented in fig. 3.6 by the range bars. In the case of a square channel with $L_C = W_C$ the transition zone TZ is symmetric. For $L_C > W_C$ the upper part of the transition zone is larger than the lower part. The opposite is the case for $L_C < W_C$. For a longer channel, percolation starts later, but probabilities rise more quickly. At the upper end the slope decreases however, meaning that it needs higher D to achieve percolation for all sample networks. For wider channels the percolation probability first rises slowly, above p_{50} probabilities however rise faster.

Before coming to conclusions, it is important to highlight that these results are only based on the existence of conductive paths in the CNN from one edge to the other. It does not say whether the conductive path is narrow or spans over the complete channel area. The next section will therefore discuss the topography of the forming networks.

3.2.2. Topography of forming networks

For a stable TFT technology it is important to have reproducible device performance. Amongst other things, this demands only low spreading in current levels for identical channel geometries. In the case of percolating

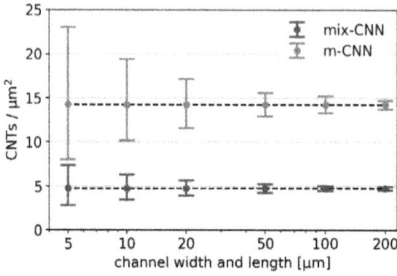

(a) square channel scaling

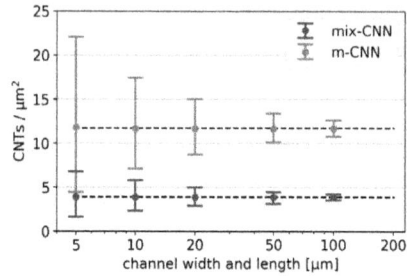

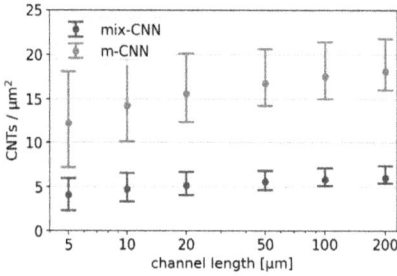

(b) channel length variation, $W_C = 10\mu m$

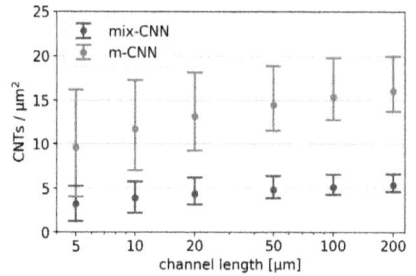

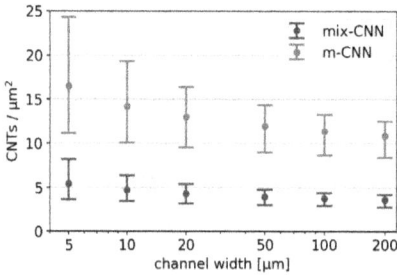

(c) channel width variation, $L_C = 10\mu m$

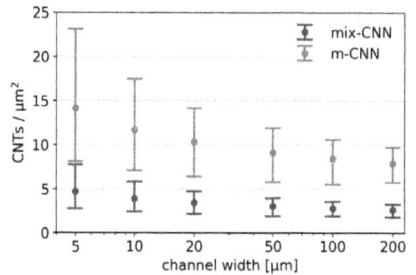

Figure 3.6.: TZ_{mix} and TZ_m for CNTs with constant length $L_t = 1$ and different channel geometries; $mc = 1/3$, $N = 1k$.

(a) square channel scaling

(b) channel length variation, $W_C = 10\mu m$

(c) channel width variation, $L_C = 10\mu m$

Figure 3.7.: TZ_{mix} and TZ_m for CNTs with length distribution as shown in fig. 3.11 and different channel geometries; $mc = 1/3$, $N = 1k$; discussion in section 3.2.3.2.

CNT networks, it is therefore important that the electrically active network is spreading over the complete channel surface to achieve a comparable density of conductive paths.

To better understand how the CNNs are forming, fig. 3.8 shows examples of CNNs for two different channel geometries. For each geometry 4 specific densities were chosen. These are from top to bottom at $D = p_{0.5mix}$ where percolation is just about to happen in the mix-CNN, p_{50mix} where only 50% of the random samples show conductive paths, $p_{99.5mix}$ where almost all samples are forming conductive paths and at the center of the semiconducting range which is defined as

$$SR_c = \frac{SR}{2} + p_{99.5mix} = \frac{p_{0.5m} - p_{99.5mix}}{2} + p_{99.5mix} = \frac{p_{99.5mix} + p_{0.05m}}{2}. \tag{3.7}$$

This would be the most stable density to always have effectively semiconducting CNNs even with metallic content while not getting too close to neither $p_{99.5mix}$ nor $p_{0.05m}$. The individual SWNTs in fig. 3.8 are shown as white sticks on black backround. All SWNTs belonging to a cluster that forms a conductive path along the channel are displayed in red.

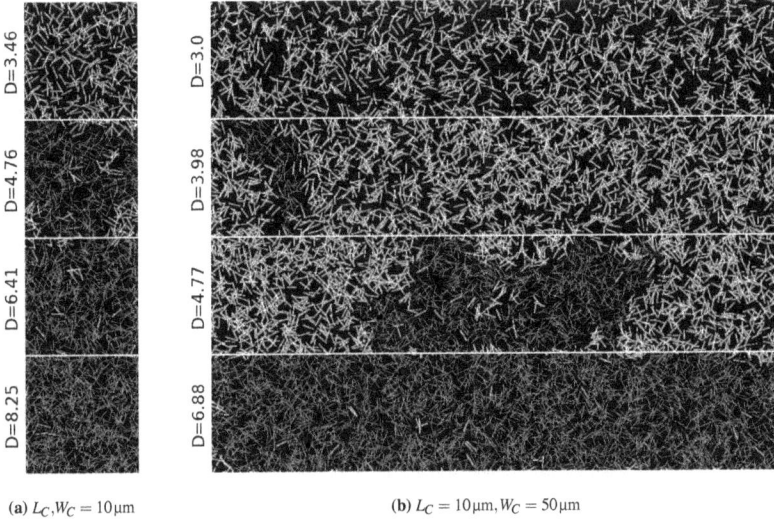

(a) $L_C, W_C = 10\,\mu m$ (b) $L_C = 10\,\mu m, W_C = 50\,\mu m$

Figure 3.8.: Topography of mix-CNN channels with D from top to bottom representing $p_{0.5mix}, p_{50mix}, p_{99.5mix}$ and SR_c which is the center of the SR; SWNTs forming electrically active networks between upper and lower edge in red.

Figure 3.8a shows random examples for the 10 μm square reference geometry. For $D = p_{50mix}$ a percolating network was chosen. It sparsely covers almost the complete channel area although at the contact edges the conductive paths would be rather confined. At $D = p_{99.5mix}$ the channel area is covered more densely and finally at $D = SR_c$ almost all SWNTs form a percolating cluster. The current through the CNN would therefore be directly dependent on the density of deposited CNTs. For this channel geometry basically all CNNs with $D > p_{99.5mix}$ cover the complete area.

In contrast to this, the examples for $W_C/L_C = 50/10$ and identical channel length show that forming paths can be quite narrow and even at $D = p_{99.5mix}$ sometimes only small sections of the deposited CNTs contribute to the conductive path (see fig. 3.8b). At SR_c the channel is however also completely covered.

For homogeneity and reproducibility reasons an important figure of merit is therefore the ratio of electrically active (connected) to deposited nanotubes which is further called **CNT active-ratio**. Figure 3.9 shows the detailed percolation probability curves on which fig. 3.6 is based. In addition box and whisker plots are overlayed that represent the CNT active-ratio for $N = 100$ random networks per density. This gives a fast overview of the value range for each geometry and network density. Only the combination of percolation probability and active-ratio sheds light on the quality of a CNN for TFT applications. To allow 100% functionality the CNT density needs to be somewhere in the SR and for a good reproducibility the active-ratio should at the same time be as close as possible to 1.

When looking at square channels (see fig. 3.9a) from $10\,\mu m$ edge length and above the CNT active-ratio at SR_c (indicated by the dashed vertical line) is always above 90%. The small channels with $5\,\mu m$ edge length exhibit however some outliers with down to 70% active-ratio and even above $p_{0.5m}$ there are some samples with rather low active-ratio. This fact combined with a rather narrow SR makes the $5\,\mu m$ square channel an unfavorable geometry. Obviously the bigger the square channels become, the more stable the results are due to a high active-ratio and large SR.

In the case of a constant channel width of $W_C = 10\,\mu m$ and varying channel length (see fig. 3.9b), this trend is even more pronounced. The CNN with $L_C = 5\,\mu m$ has even lower active-ratios. Evidently for $L_C = 10\,\mu m$ the results are identical with fig. 3.9a as it is also a square channel. For longer channels SR is getting even larger and the active-ratios are close to 100% already at $p_{99.5mix}$.

For channels with constant $L_C = 10\,\mu m$ and changing width the transition zone for the CNT active-ratio shifts towards higher CNN densities for increasing W_C/L_C ratio. At the extreme case of $W_C = 200\,\mu m$ even at $p_{99.5mix}$ the CNT active-ratio is only around 10%. Nevertheless in all shown cases the active-ratio at SR_c is above 90% and approaches 100% close to $p_{0.5m}$.

In summary it can be said that a low $W_C/L_C < 1$ is beneficial for a stable CNT-TFT technology. It leads to a CNT active-ratio close to 100% and even for classical mixed CNT feedstock with 1/3 metallic content a large effectively semiconducting range is achieved. For $W_C/L_C > 1$ and $L_C/L_t \leq 10$ stable results can be achieved as well, but the control over network density needs to be much more precise.

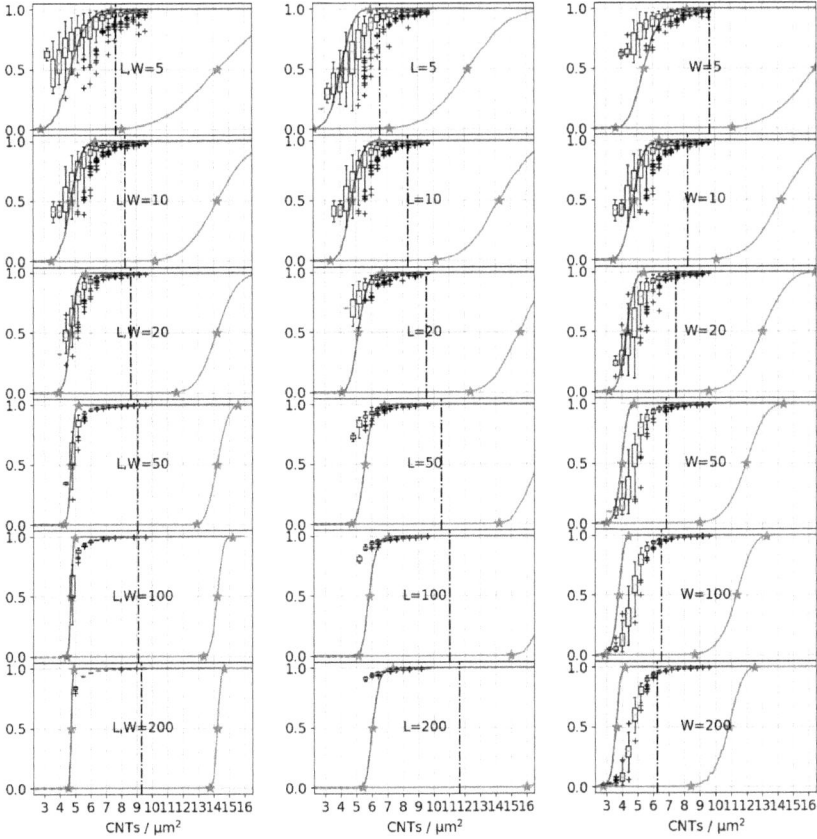

(a) Square scaling, $L = W$; boxplot data for $L,W = 200\,\mu m$ based on 10 simulation runs per density.

(b) Length variation, $W = 10\,\mu m$

(c) Width variation, $L = 10\,\mu m$

Figure 3.9.: Percolation probability of mix-CNN (blue) and met-CNN (red) for different channel geometries; stars indicate positions of $p_{0.5}$, p_{50} and $p_{99.5}$; vertical dashed line is center of SR; box and whisker plot shows ratio of electrically active vs. deposited CNTs from 100 random CNNs per density (red line: median, box: upper and lower quartille, whiskers: upper and lower limit, pluses: outliers); $L_t = 1$, $mc = 1/3$, $N = 1k$.

3.2.3. Influence of carbon nanotube length

In the previous section the influence of the channel geometry while keeping a constant CNT length was described. In this section the focus is put on varying the CNT length.

3.2.3.1. Uniform CNT length

The influence of percolation probability for an identical square channel but different L_t is shown in fig. 3.10. Evidently shorter CNTs demand a higher density to achieve percolation. The double logarithmic plot in fig. 3.10a shows an almost perfect linear dependency between D and L_t. Shorting L_t at constant channel

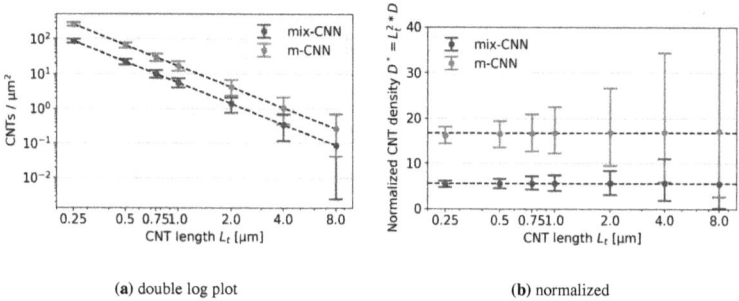

(a) double log plot   (b) normalized

Figure 3.10.: Influence of CNT length on percolation probability; $W_C = L_C = 10\,\mu m$, $L_t = 1\,\mu m$, $M = 800$, $mc = 1/3$, $N = 1k$.

geometry has basically the same effect as increasing the channel geometry for constant CNT length (compare fig. 3.6a). This becomes more evident when normalizing D for each CNT length category to the same L_t/W_C ratio as the reference ($L_{tref} = 1\,\mu m$, $W_C = L_C = 10\,\mu m$).

$$D^* = \frac{L_t^2}{L_{tref}^2}D = L_t^2 \cdot D \cdot 1\,\mu m^{-2} \qquad (3.8)$$

The resulting plot is given in fig. 3.10b. To limit non-linearities due to quantization from the rastering $M = 800$ was chosen. This results in slightly higher p_{50} levels as in fig. 3.6a.

The normalized representation of the data indicates also clearly that using shorter CNTs would narrow down the transition zones and result in a large effectively semiconducting range. So instead of scaling up channel geometries, shorting of the CNTs would give the same results while keeping a small footprint. In both cases the result is however a shorter intra-tube path length and more inter-tube transitions for charge carriers which limits the conductance of the CNN as discussed in section 2.1.4.

Another crucial assessment can be derived from fig. 3.10b. At a CNT length of $L_t = 2\,\mu m$ the semiconducting range SR has almost vanished. It can therefore be postulated that the ratio between channel length and nanotube length L_C/L_t should be at least 5:1 for having a chance to achieve an effectively semiconducting network with uniform length CNTs. When looking at figs. 3.6a and 3.6b it becomes clear that this somehow also depends on the width of the channel. The postulated factor of 5:1 seems however to be a good first order estimate.

3.2.3.2. CNT length distribution

In reality SWNT feedstock doesn't deliver tubes of identical length but a distribution of different CNT lengths. Unfortunately this length distribution is often not well known and processes used for dispersing of the nanotubes (see section 4.4) can further modify the CNT lengths. The s-SWNT enriched IsoNanotubes-S from NanoIntegris used for CNN-TFTs in this work are already delivered as suspension. Further the CNT length distribution, characterized by atomic force microscopy, is approximatively specified in the product data sheet. Although this specific length distribution is probably not representative for other SWNT suspensions used in this work it will be used here to show the effect on CNN percolation behavior. Especially as the IsoNanotubes-S dispersion gives best results in CNN-TFTs.

The IsoNanotubes-S length distribution specified in the product's technical data sheet is given in fig. 3.11 by the red line. For the following Monte-Carlo simulations this distribution is approximated by using a beta distribution [22, 136]. It draws samples between 0 and 1. The form of the distribution can be modified by the two shape parameters α and β. A good fit of the envelope is achieved with $\alpha = 1.7$ and $\beta = 18$. Multiplied by a factor of 10 and with an added offset of 0.16, a good correlation can be achieved as shown by the histogram of 5 million CNT length samples shown as blue bar chart in fig. 3.11. The maximum length is limited to $L_t \leq 3\,\mu m$.

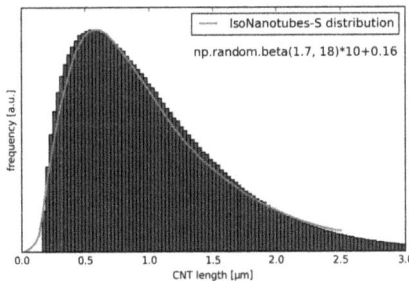

Figure 3.11.: Length distribution of IsoNanotubes-S and histogram of $5 \cdot 10^6$ CNTs approximated by beta distribution up to max $3\,\mu m$ length.

The same simulations as discussed in sections 3.2.1 and 3.2.2 were repeated. The only difference being the CNT length distribution instead of a constant length. It needs to be emphasized that the mean CNT length of the distribution stays at $\overline{L_t} = 1\,\mu m$. A direct comparison of the changing characteristics is therefore possible. For easier comparison the results are shown in fig. 3.7 in section 3.2.1. The detailed plots of percolation probability and CNT active-ratio can be found in fig. C.1 in appendix C.

For all channel geometries the percolation probabilities shift towards lower network densities and the transition zones TZ_{mix} and TZ_m get larger. As a consequence the semiconducting range becomes smaller. For the given length distribution the above postulated factor of $L_C/L_t > 5$ is not valid any more and needs to be redefined to $L_C/\overline{L_t} > 10$. The CNT active-ratio also slightly decreases at the center of the semiconducting range SR_c.

3.2.4. Influence of metallic to semiconducting ratio

All simulation results discussed so far were done for the standard mix with a metallic content $mc = 1/3$. High concentration semiconducting feedstock is however available nowadays. In section 3.1.2 the relation of $p_{50m} = 3 \cdot p_{50mix}$ was already stated for the standard mix. This becomes evident when looking at how the simulation is done. Two lists of random CNT coordinates are created. The number of tubes reflects the defined ratio. With a sufficiently high number of samples the m-CNN will behave identical to the mix-CNN with the only difference that the corresponding percolation densities are three times higher as for the mix-CNN. It therefore holds that

$$\frac{p_{Xm}}{p_{Xmix}} = \frac{TZ_m}{TZ_{mix}} = \frac{1}{mc} \tag{3.9}$$

where the subscript "X" stands for any percolation probability. This is further also true for any length distribution. As an example table 3.1 lists percolation probabilities for 98 % s-SWNTs ($mc = 0.02$). Values for the mix-CNN come from simulation results shown above, the m-CNN values are calculated with eq. (3.9). The listed values show that in the case of the β-distribution the semiconducting range significantly diminishes compared to $L_t = 1\,\mu m$. In this special case SR decreased to 74 % of the SR of $L_t = 1\,\mu m$.

Table 3.1.: Percolation probabilities for $mc = 0.02$ and $L_C = W_C = 50\,\mu m$ for CNTs with uniform length and beta distribution; mix-CNN values simulated, m-CNN values calculated with eq. (3.9); all values in [CNT/μm^2].

L_t	mix-CNN				m-CNN				SR
	$p_{0.5}$	p_{50}	$p_{99.5}$	TZ_{mix}	$p_{0.5}$	p_{50}	$p_{99.5}$	TZ_m	
1 μm	4.3	4.8	5.3	1.0	213.0	237.6	262.5	49.6	207.7
β-distribution fig. 3.11	3.2	3.9	4.4	1.3	157.9	194.7	222.5	64.6	153.4

4. Carbon nanotube feedstock and preparation methods

The methods used for CNT synthesis and the following treatments for purification and optionally sorting have a big influence on price and quality of the feedstock. Depending on the application, one material might be preferred over another. For the devices fabricated in this work SWNT and DWNT feedstock was purchased in form of powder and dispersed for deposition from wet phase. In some cases SWNTs were already purchased as dispersion. The synthesis, purification and sorting of SWNTs is not part of this work. For better understanding CNT synthesis and treatment methods are however also quickly discussed in this chapter. The corresponding sections 4.1 and 4.2 are taken from [118] which is a contribution to the Handbook of Visual Display Technology [19] by the author of this thesis himself. At the end of this chapter, the used CNT feedstock and the preparation of SWNT dispersions from powder feedstock is presented.

4.1. Carbon nanotube synthesis

There are three primary methods to produce carbon nanotubes: arc discharge, laser ablation, and chemical vapor deposition. In all three cases, transition metal catalysts like Fe, Co, Mo, Ni, or Y are necessary to get mostly single-walled CNTs. The relatively simple and cheap setup of the arc-discharge method enabled lots of groups worldwide to produce their own CNTs leading to a huge increase in nanotube research. Between two carbon electrodes, a DC plasma is generated by an arc discharge in a certain gas atmosphere. The anode, containing small amounts of the metal catalyst, is being consumed, while the nanotubes condense at the cathode [99]. In the laser-ablation method, the carbon target including the metal catalyst is evaporated by a high power laser in a furnace under 1200 °C in an inert gas flow. The nanotubes are collected at a cooled surface outside the furnace [130]. With both methods the nanotubes are collected as powder with purities of 70-90%, the rest being amorphous carbon or other unwanted carbon molecules and catalyst clusters. A controlled synthesis on a substrate surface or a directed deposition is not possible. Due to strong van der Waals forces, the nanotubes form ropes or bundles consisting of several tens to hundreds of single tubes. The formation of these bundles is a major issue for subsequent processing since in most applications and specifically if used as semiconductor in TFTs, well-separated individual SWNTs are preferred. Mass production with both techniques seems unlikely because of high power-consumption and scaling issues.

In the CVD approach, some kind of gaseous carbon feedstock is decomposed under temperatures from 700 °C to 1200 °C. Typical carbon sources are hydrocarbons like methane (CH_4) or carbon monoxide (CO). The metal catalysts in the form of nanoparticles can be predeposited on the substrate, or they might be formed during the CVD, for example, by the decomposition of an organometallic species like in the HiPCO process [15, 92]. HiPCO stands for high-pressure catalytic decomposition of carbon monoxide and is a promising candidate when it comes to high-volume mass production. With various choices in carbon source, catalyst, temperature, gas atmosphere, and reactor design, there are many flavors of CVD synthesis. So far the most promising besides HiPCO are methane CVD, CO CVD, alcohol CVD, and plasma-enhanced CVD (PECVD) [65].

Advantages of the CVD processes are the possibility of direct deposition or growth of well-separated SWNTs on the substrate. By choosing the right parameters and substrates, alignment and position selective growth can be achieved. The drawback however is the limited choice of substrate materials due to the high reactor temperatures. Successful growth at lower temperatures was reported by several groups. The lower temperatures are however also a cause for higher impurity or defect rates. Using the advantages of direct growth on display grade glass or even plastic substrates seems so far unfeasible. Transfer techniques [124] have proven to be a viable way for applying CVD-grown nanotube arrays on plastic substrates.

A highly appreciated milestone in CNT synthesis would be the growth of a single (n, m) species or at least of a certain electronic type. By varying synthesis parameters, catalyst material, and carbon feedstock, researchers were able to get better control over the SWNT diameter [50]. Since the number of possible geometries decreases with smaller diameters (<1.2 nm), samples with narrow distributions of (n, m) were realized [30, 87]. With certain growth parameters, several groups achieved the preferential synthesis of semiconducting [42, 83, 84, 102] or metallic [40, 134, 143] SWNTs. Reported purities lie in the range of 90 % to 95 %. Just recently the synthesis of >95 % s-SWNT combined with a high nanotube density was reported [67]. These results are a big improvement compared to the 1/3 metallic to 2/3 semiconducting ratios for most of the synthesis methods.

For application in field-effect transistors, even m-SWNT portions <1 % can however limit device performance especially when going to shorter channel lengths. The synthesis of a single (n, m) type stays challenging, but recent publications reported promising results. A possible route might be the growth from template-like CNT fractures [135] or specially synthesized molecules [61, 111]. An outstanding result was presented by Wang et al. who achieved the synthesis of almost 52 % (9,8) s-SWNTs by using a special catalyst mixture. With a diameter of 1.15 nm, these nanotubes are significantly larger than most of the sub-1 nm tubes preferentially produced by the other listed methods. This is important as small diameter nanotubes tend to form higher Schottky barriers when contacted by metal layers.

4.2. Purification and sorting methods

As mentioned before purification and sorting methods are not part of the practical work of this thesis. The methods are nevertheless described to give a better understanding of the influences of such treatments.

4.2.1. Purification

With all synthesis methods not only SWNTs are produced but also significant amounts of by-products. These are usually metal catalyst particles as well as diverse carbonaceous elements like amorphous carbon, fullerenes, MWNTs, and other carbon nanoparticles [97]. In order to get high-purity samples, these impurities need to be removed. Several methods have evolved that can be divided in three categories: chemical methods like acid- or gas-phase treatments; physical methods like filtration, centrifugation, or chromatography; and multistep methods combining chemical and physical techniques [52]. A very common approach is gas-phase oxidation of impurities with higher reactivity, followed by acid reflux with typically nitric acid [24]. While some species like amorphous carbon, fullerenes, or metal particles can be removed with high selectivity due to higher reactivity or solubility in certain media, some carbonaceous molecules have almost identical chemical reactivity. This means that there is always a trade-off between purity and damaging of the nanotubes. Since every synthesis method leads to different mixtures, as-produced purities, and mean CNT diameters, the purification technique needs to be tailored to the synthesis method.

4.2.2. Sorting by electronic type

Since nanotubes are always synthesized in a variety of geometries with different lengths, diameters, and electronic type, diverse sorting techniques were explored in the past. While sorting by length [4, 56] or diameter [5] can be beneficial for certain processing techniques, the biggest interest lies in sorting by electronic type especially since the selective synthesis still needs much more research. For many sorting techniques, a starting material with low deviation in diameter or length is however advantageous or mandatory. Since metallic and semiconducting nanotubes are well mixed over all kinds of diameters and chirality, finding some kind of selective process concerning the electronic type took many years. The first reported enrichment in metallic SWNTs was in 2003 [77], 10 years after synthesizing the first SWNTs. As with synthesis, purification, and dispersing of nanotubes, diverse post-synthesis methods are pursued in order to get carbon nanotube material of only one electronic type. The review of Hersam gives a comprehensive overview [50]. The most promising approaches with regard to TFT applications are presented in the following sections categorized in selective elimination and nondestructive sorting [65].

4.2.2.1. Selective Elimination

Especially when building transistor devices with directly grown nanotubes, the only way of avoiding metallic shortcuts is by selectively eliminating them after the deposition. One method that was used early on is electrical breakdown of the unwanted m-SWNT [20]. The s-SWNTs are rendered nonconducting by a high gate field in the cutoff region, leaving only the metallic nanotubes for electrical conduction between the source and drain contacts. High DC or pulsed currents are sent through the TFT channel that leads to thermal decomposition of the conductive nanotubes [120]. A similar approach was used by Kern et al. where the s-SWNTs are turned off in the same way, while the resistivity of the m-SWNTs is increased significantly by an electrochemical modification [8]. There are also reactions that do not rely on the depletion of the s-SWNTs and electrical contacting, making them more suitable for large-area applications. Hassanien et al. demonstrated that a hydrogen plasma treatment favors etching of metallic nanotubes over semiconducting ones [41]. Dai et al. presented 100 % yield of s-SWNTs using a methane plasma followed by an annealing process leading to selective hydrocarbonation of m-SWNTs [145]. The treatment works for SWNT diameters from 1.3 nm to 1.6 nm. For lower diameters, both types are being altered owing to the higher reactivity caused by the increased strain in the C-C bonds. For larger diameters, the metallic nanotubes stay intact. There are also optically driven approaches like the photolysis-assisted oxidation of m-SWNTs by laser irradiation in air [54]. Recently, also microwave radiation was demonstrated to introduce thermocapillary effects in m-SWNTs that can be used to realize large-area CVD-grown s-SWNT arrays [139]. In all processes with selective elimination, care has to be taken not to damage the wanted species especially since the selectivity often is not very high. The effectivity of these methods is usually improved by having well-aligned arrays with limited nanotube density which on the other side is not in favor of miniaturization.

4.2.2.2. Nondestructive Solution-Based Sorting

The first presented approach for enrichment in electronic type was by dielectrophoresis [77]. Between microelectrodes, an AC electric field is applied. The nanotubes suspended in an aqueous surfactant solution floating over the electrodes are polarized in the electric field and are dragged toward higher field intensity [25]. Because of differences in their dielectric constants, the effective force is unequal for m-SWNTs and s-SWNTs. Owing to their higher polarizability, m-SWNTs generally feel a higher force and show higher probability for deposition between the electrodes. For high frequencies, the force on s-SWNTs is even reversed [76]. By closely controlling all of the parameters, m-SWNT can be deposited exclusively, leading to enrichment of s-SWNTs in the suspension. The method is however strongly depending on the geometrical properties of the nanotubes. Length and diameter should therefore be almost identical. Also bundles should be avoided, since the effective force scales proportional to the length and the square of the diameter. Scaling of this method seems difficult, and a complete separation was not achieved so far. Using a microfluidic channel, Shin et al. were able to extract highly metallic samples with good prospects for scalability [123]. The selective deposition of CNNs in the TFT channel from mostly semiconducting feedstock by dielectrophoresis was tested in this work. A sufficiently high performance was however not achieved as described in chapter 7.

The breakthrough in electronic-type sorting was achieved by using ultracentrifugation in a density gradient medium (density gradient ultracentrifugation, DGU) and led to commercialization of highly enriched SWNT dispersions in 2007. The technique allows to sort the species injected into the density gradient medium by its buoyant density during ultracentrifugation. It was first used to sort SWNTs by their diameter [5]. Using mixtures of two surfactants, Hersam et al. managed to separate SWNTs by their electronic type with purities up to 99 % and more [6]. The selectivity comes from the differences in polarizability of m-SWNT and s-SWNT and the two surfactants that competitively adsorb to the SWNT surface, leading to different buoyant densities for different species. In order to receive high-purity samples, several iterations and starting material with a limited range of (n, m) as achieved by the CoMoCAT method [7] are necessary. Ghosh et al. reported a drastically improved efficiency by using tailored nonlinear density gradients. With this improvement, they were able to sort highly polydisperse HiPCO material into individual (n, m) fractions in a single step (see fig. 4.1a) [35]. The DGU method stays however a batch process with limited scalability. In the meantime,

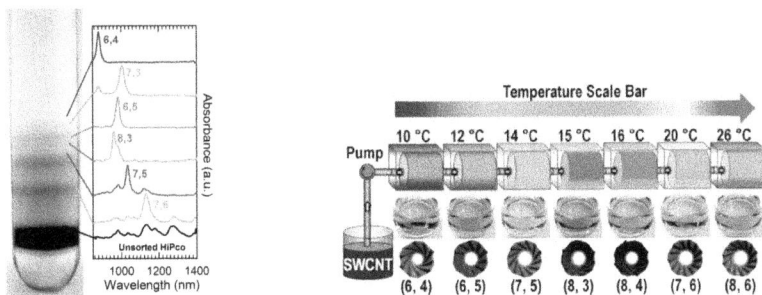

(a) Ultracentrifugation in a nonlinear density gradient medium; image of a centrifuge tube containing HiPCO SWNTs sorted by one 18-h run at 268000 g. The distinct colored bands are layers enriched in different SWNT species. The near-infrared absorbance spectra of the marked colored layers show the main (n, m) component. Spectra are normalized and offset for clarity. The unsorted HiPCO spectrum is scaled down by a factor of 10 (Reprinted by permission from Macmillan Publishers Ltd: Nature Nanotechnology [35], Copyright 2010).

(b) Temperaturecontrolled gel chromatography for the high-efficiency single-chirality separation of single-walled carbon nanotubes (Reprinted with permission from [86]. Copyright (2013) American Chemical Society).

Figure 4.1.: Sorting of SWNTs by electronic type.

a variety of effective sorting methods from SWNT dispersions have been demonstrated. One approach with better scalability and perspective for high-volume, lower-cost sorting uses gel chromatography. Liu et al. further improved the sorting by additionally exploiting the temperature dependent adsorbability of SDS-wrapped SWNTs onto the gel. By using consecutive gel columns at increasing temperatures (see fig. 4.1b), they achieved separation of several distinct s-SWNT types [86]. A further technique with better prospects for upscaling is the polymer-aided dispersion of carbon nanotubes. Lately, several types of specifically chosen polymers were reported to selectively disperse s-SWNTs in organic solvents like toluene [80, 91, 116, 132]. These polymers wrap around the nanotubes with the help of sonication. The dispersions are then cleaned from impurities by ultracentrifugation, identical to the surfactant-aided aqueous dispersions (compare section 4.4). While most of the sorting techniques work best with smaller diameter SWNTs, Brady et al. presented polymer-aided dispersions up to a diameter of 1.8 nm [13].

4.3. Utilised CNT feedstock

In the course of time different carbon nanotube feedstock was purchased and tested. Table 4.1 summarizes the different types of carbon nanotubes that were used in this research project. All nanotube types besides IsoNanotubes-S were characterized for their use as transparent conductive films. The arc-discharge nanotubes gave best performance for the TFT application with the standard mix of s-SWNTs and m-SWNTs. Later on the highly semiconducting IsoNanotubes-S were used.

Table 4.1.: Different types of utilized carbon nanotube feedstock.

Name	Vendor	Synthesis	delivered as	Impurities [wt%]	mc	sc	diameter [nm]	length [μm]
Arc-discharge		Arc discharge	powder		0.33			
HiPCO ap	Unidym	HiPCO	powder	< 35	0.33		0.8-1.2	0.1-1
HiPCO p	Unidym	HiPCO	powder	< 15	0.33		0.8-1.2	0.1-1
HiPCO sp	Unidym	HiPCO	powder	< 5	0.33		0.8-1.2	0.1-1
DWNT	Unidym	HiPCO	powder					
IsoNanotubes-M	NanoIntegris	Arc discharge	pellet	< 5	0.95		1.4 (1.2-1.7)	0.5 (0.1-4)
IsoNanotubes-S	NanoIntegris	Arc discharge	suspension	< 5		0.98	1.4 (1.2-1.7)	1 (0.1-4)

4.4. Preparation of CNT dispersions

All deposition methods used in this work rely on working with a liquid phase. Single and double walled carbon nanotubes are purchased in powder form. The best working medium that was found in a former work uses aqueous surfactant stabilized suspensions [117]. The surfactant molecules form micelles around the CNTs in a way that the non-polar tails of the surfactant molecules go in contact with the also non-polar CNTs. The polar heads of the surfactant molecules are turned outwards. Individual micelles stay at distance of each other due to homopolar electrostatic charges. This phenomenon stabilizes the suspension and keeps it longtime stable. The preparation of suspensions is as follows.

First a surfactant solution of 10 weight percent (wt%) sodium-dodecyl-sulfate (SDS) or lithium-dodecyl-sulfate (LiDS) in de-ionized water is prepared. The SWNT powder is weighed into a glass bottle with a high precision scale (fig. 4.2a). Then the surfactant solution is added to achieve a ratio of 0.1wt% SWNTs in surfactant solution. For fine dispersion of individual nanotubes, a high intensity sonication with a sonotrode that reaches directly into the mixture is necessary (fig. 4.2b). Simple mixing, stirring or sonication in a bath proved not to be effective enough to achieve significant dispersion. [117].

The duration of the sonication treatment depends on the dispersibility of the raw nanotube feedstock. Different synthetisation methods and following purification processes lead to quite different density and texture of source material. A sufficiently long treatment is necessary to effectively exfoliate the nanotube bundles that form with high inter-tube Van-der-Waals forces. A high ultrasound dosage can however also damage or cut the SWNTs which is of disadvantage for the discussed applications. Due to a lack of characterization methods for the individual tubes and bundles in the suspension, an optimum duration could not be determined. The standard treatment was 1 h at full sonotrode power. This gives a deep black liquid that is long time stable over years (fig. 4.2c).

The so far prepared suspension still contains larger particles and nanotube bundles that could not be exfoliated. Ultracentrifugation is used for further purifying the nanotube suspension. At acceleration forces of 40 000 g, the suspension is centrifuged for 2h at room temperature. This sorts the suspended particles by size, concentrating larger bundles and other impurities that tend to precipitate at the bottom of the vessel. The purified suspension is afterwards carefully decanted while keeping the lower particle swamp in the vessel. The result is a homogeneously dispersed suspension that is long-time stable and can be further diluted if needed (fig. 4.2d).

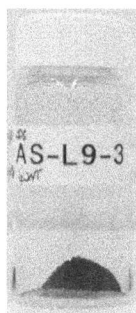

(a) SWNT powder   (b) Dispersing in surfactant solution with help of sonotrode.   (c) Suspension after sonication   (d) Centrifuged and diluted suspension

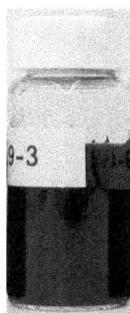

Figure 4.2.: SWNT suspension preparation

5. Deposition and patterning of carbon nanotubes on glass and plastic substrates

Fundamental research on individual carbon nanotubes and CNT networks has revealed their extraordinary performance quite soon after their discovery. One of the great challenges that need to be overcome before CNTs can be used for production processes is their handling. The nanoparticle nature offers new characteristics, it imposes however also great challenges. Ideally, placement and orientation of each individual nanotube should be controlled while using fast, scalable and cost-effective methods. For the application as transparent conductive films, it is quite obvious that some kind of a network needs to be created to realize conductive areas. Single tube devices are imaginable in the case of CNT transistors. While these are important for basic research, they show however limited practical use for the applications discussed in this work because the current flow through individual SWNTs is too low. For homogeneity reasons it would further be necessary to have identical tubes in all devices which is still not possible even with advanced growing and sorting methods. Further, the realization of a deposition method that allows full control over placement and orientation of each individual tube with high yield and repeatability seems so far out of reach.

Not all applications demand however such a high level of control during deposition. The method of choice also depends strongly on the intended use. The different methods will therefore be classified in the two applications that are dealt with in this work - transparent conductors and the semi-conducting layer in TFTs. The substrate preparation applies however for both types of applications and is described below. Further deposition methods known from literature but not used in this work are listed at the end of this chapter to give a more complete picture of possibilities.

Throughout this work, two types of substrate materials are used: Corning® display grade Eagle 2000™ glass as well as a polyethersulfone (PES) based foil from Sumitomo Bakelite Co., Ltd. for flexible applications.

5.1. Substrate preparation

Carbon nanotubes can be deposited on various surfaces. Glass and non-birefringent plastic substrates are the materials of choice for display applications and especially for liquid crystal displays. Although the

CNNs can be deposited without surface treatment, the following processing might demand adhesion improvement.

5.1.1. Etch-stop and metal adhesion layer on plastic substrates

Working on plastic substrates is much more delicate than on the well established glass. This is due to a much lower glass-transition temperature T_g, which is strongly limiting the process temperature range, but also due to a lower mechanical and chemical stability. Especially certain solvents and plasma treatments attack the surface much more easily. The CNNs in this work are often patterned by reactive ion etching with an O_2-plasma which also attacks the plastic surface (see section 5.6.2). Inorganic layers, that are not or only slowly etched by the plasma, can be applied prior to CNN deposition, thus preventing trenches in the substrate. In addition, the same layer can serve as adhesion promoter for metals that sometimes have poor adhesion on plastic substrates.

In this work, several thin inorganic layers (SiN, SiO_2 and Ta_2O_5) deposited on the plastic substrate by sputtering were tested for their applicability as etch-stop layer. Important factors are high transparency, good adhesion to the plastic surface, etch rate in the dry etch process and adhesion of the CNNs on the etch-stop layer. Thin layers of 30 nm thickness are used to keep the layers as flexible as possible. The influence of such thin layers on the optical transmission is very low.

The qualitative results are listed in table 5.1. Overall Ta_2O_5 gives the best performance. It can be deposited with a fast rate, is not etched during the RIE step and scotch tape tests resulted in good adhesion on the PES substrate. The CNN shows also sufficient adhesion during the scotch tape test when covered with an adhesion layer for nanotubes (see section 5.1.2). A 30 nm thick etch-stop layer of Ta_2O_5 was therefore chosen as standard when working with patterned CNNs on plastic.

Table 5.1.: Qualitative summary of tested etch-stop and adhesion layers.

Material	deposition rate	O_2 RIE etch rate	Adhesion to PES	Adhesion of CNN
SiN	–	$10\,\mathrm{nm/min}$	+	+
SiO_2	o	0	+	+
Ta_2O_5	+	0 [1]	++	+

5.1.2. Self-assembled adhesion layer for CNNs on inorganic surfaces

A good adhesion of the nanotubes to the substrate surface is mandatory for its usability in devices. The adhesion on organic surfaces like plastic substrates has proven to be sufficient for wet processing after deposition. The prior section shows however that other process requirements necessitate additional layers on the plastic substrate. Larger flakes of the CNN can detach from inorganic surfaces like glass substrates

---

[1] peaks up to 12 nm forming after several minutes of etching

or oxides used as etch-stop layer if there is too much agitation during rinsing in water. An amine-terminated silane is deposited as self-assembled monolayer (SAM) to improve the adhesion to such surfaces [89, 94, 119]. This is done by preparing a 1 % solution by volume of (3-Aminopropyl)triethoxysilane (APTS) (see chemical structure in fig. 5.1a) in de-ionized water and soaking the substrate in it for 1 h. The three ethoxy groups (CH_3CH_2O) form covalent bonds with the OH groups at the surface, resulting in a well attached monolayer with an amine-terminated tail (see fig. 5.1b). The substrates are then rinsed in DI-water to remove excessive non-bonded silane solution and blown dry with N_2. This SAM is hydrophobic, which is good for the adhesion of the nanotubes, it impedes however spreading of aqueous solutions. Such monolayers are reported to be longtime stable. Usually, the treatment is applied shortly before nanotube deposition to assure identical conditions for each deposition.

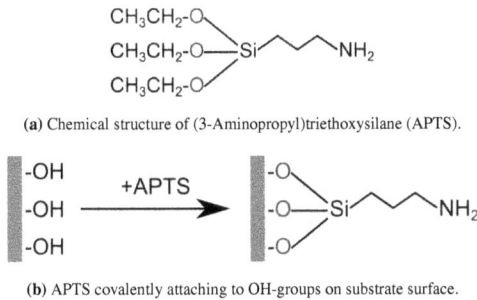

(a) Chemical structure of (3-Aminopropyl)triethoxysilane (APTS).

(b) APTS covalently attaching to OH-groups on substrate surface.

Figure 5.1.: Silanization of inorganic surfaces leading to amine-terminated self-assembled monolayer.

5.2. Deposition method for use as transparent conductor

To achieve a laterally isotropic conducting layer with a high transmission from rod-like molecules, the ideal formation is a randomly oriented pseudo-2D network. A relatively high nanotube density is necessary to achieve a sufficiently low sheet resistance in the range of some $10\,\Omega/\square$ to some $1000\,\Omega/\square$. This can be achieved with a spray coating approach.

In this work a simple off-the-shelf airbrush pistol was used for the deposition (see fig. 5.2a). The general setup of the nozzle is shown in fig. 5.2b. Compressed air (or in this case pressurized N_2 for purity reasons) is guided along a nozzle containing the spray medium. As described in section 4.4, it is a surfactant based SWNT dispersion. A fine needle is placed in the opening of the nozzle for controlling the material flow. When the needle is drawn back the N_2 flow pulls out the spray medium by the Venturi effect. The liquid moves along the needle and is atomized at the tip. The material flow can be adjusted by regulating the position of the needle. Further parameters are gas pressure, distance to the surface and the nozzle diameter.

The fine atomized spray hits the surface, where the water content evaporates. Bigger droplets form on the substrate before the water can evaporate, when the spray flow is too high. The free surface energy of the substrate is usually too small to lead to full spreading. Therefore, growing droplets contract and lead to an inhomogeneous deposition. This can be avoided by placing the substrate on a heated surface of 50 °C to 100 °C. For plastic substrates, a heated vacuum chuck should be used to avoid a buckling substrate and resulting inhomogeneous temperature dissipation.

The airbrush forms a cone of atomized spray with a limited area coverage at practical distances of 20 cm to 30 cm between nozzle and substrate. For a homogeneous deposition, a meander is followed (see fig. 5.2c). Rotating the substrate by 90° between passes further helps to settle lateral differences in nanotube coverage. Once a desired network density is achieved, which is best controlled by the sprayed suspension volume, the substrates are rinsed in water to remove the surfactant from the nanotube layer as described in section 5.5.

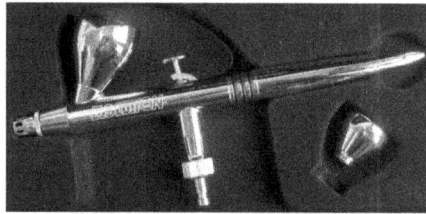

(a) Airbrush pistol used for deposition.

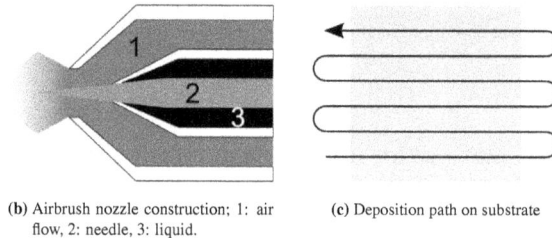

(b) Airbrush nozzle construction; 1: air flow, 2: needle, 3: liquid.

(c) Deposition path on substrate

Figure 5.2.: Spray-coating

5.3. Deposition method for use as semiconductor in TFTs

The encapsulating micelles prohibit direct contact between CNT and substrate surface when spin-coating surfactant based suspensions of carbon nanotubes. Also, the surface treatment with APTS leads to coverage with hydrophobic end-groups. The surfactant molecules quickly cover this surface with their non-polar tail. The interface between substrate and CNT micelles is therefore identical to the repelling interface between CNT micelles. Without further measures, there is basically no deposition of nanotubes on the surface. The spin-coating method used in this work was first published by the John Rogers group of the University of Illinois [89]. An organic solvent is used to remove the surfactant micelles during the deposition. This is done by mixing a stream of nanotube suspension with a stream of organic solvent right above the spinning substrate. Figure 5.3a shows the general setup with CNT suspension, organic solvent and the mixing streams just above the surface of the substrate. First trials were made with two syringes held by hand and pressurized

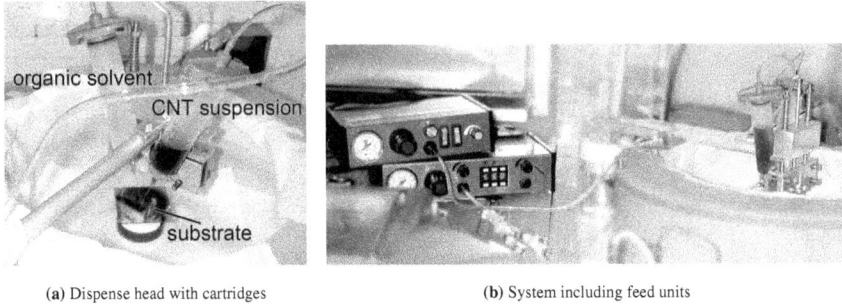

(a) Dispense head with cartridges

(b) System including feed units

Figure 5.3.: Dispensing system used for SWNT deposition by spin-coating.

with the thumbs at the same time. For a better process control an adjustable fixation as shown in fig. 5.3a was realized. This allows identical positioning of the cartridges and therefore mixing point for each deposition. The following parameters are varied and need to be kept as repeatable as possible, to find the optimal deposition:

- deposition time (total volume)
- volume flow of the individual streams (CNT suspension & solvent)
- type of organic solvent
- distance of the mixing point in reference to the substrate surface
- spinning speed of the substrate.

The control over the volume streams is achieved by using two feed units. These individually pressurize the cartridges containing suspension and solvent. The two units allow individual control over dispense time and gas pressure. The volume stream is further specified by choosing a dispense tip diameter allowing more or less volume flow. The system containing the two dispensing units is shown in fig. 5.3b. The dispensing duration of the nanotube suspension is controlled by a programmable timer. The solvent flow is established latest at the beginning of the CNT dispensing and continues a few seconds after the programmed time. The intermixing stream is likely to be dominated by one of the two streams leading to a transversal mixed stream, when using different dispense tip diameters and pressure settings. In this case the pressure in one or both units is adapted to have a mixed stream that continues vertically after the mixing point.

5.4. Further deposition techniques from literature

The so far presented deposition methods that are used in this thesis create randomly oriented networks. For the TFT application, a perfectly aligned array of purely semicondcuting nanotubes, bridging the gap between source and drain, would be an improvement as the limiting inter-tube connections could be minimized. There are several approaches using either direct growth or solution deposition methods. Very good results can be achieved with the direct growth on special cut quartz wafers where the interaction with the wafer lattice creates not only alignment but in some cases also growth of mostly semiconducting nanotubes. Transfer of such aligned arrays from the growth substrate to a display backplane with high yield adds however a new significant challenge. Several review articles concentrate on the different deposition methods for aligned arrays [55, 65, 118, 142]. At the current stage, most of these methods seem however too expensive or too experimental for the applications discussed in this work.

5.5. Surfactant removal from deposited CNN

The as-deposited CNNs have large amounts of surfactant in the network. It is mandatory to remove the surfactant molecules that enwrap the SWNTs, to achieve a good intertube contact. The simplest way to achieve this is rinsing in water. The residual surfactant dissolves fast in water, while the non-polar SWNTs prefer to adhere to each other or to the substrate surface. However, there is a risk that SWNTs are removed in micelles as long as the surfactant concentration stays high enough at the interface between water and CNN. The substrates are therefore first immersed into clean DI-water without further agitation or ultrasound. After 10 min, when the surfactant concentration is already strongly reduced, the substrates are put into a laminar flow of DI-water with the flow direction along the CNN surface for another 10 min. The samples are then blow-dried with nitrogen. Residual water can be removed by oven drying or in a vacuum chamber.

From literature it is known that nitric acid (HNO_3) treatment can further remove residual surfactants and additionally densify the CNN [34]. HNO_3 or alternatively thionyl chloride ($SOCl_2$) also acts as dopant, increasing the free charge carrier density and hence giving better conductivity (see also section 2.1.4). This doping effect is however volatile without further measures. Such treatments were not applied in this work because the chemicals are toxic and might further damage existing layers like metals due to their acidic nature.

5.6. Patterning of CNN layers

Generally speaking there are two fundamental approaches to achieve patterned CNN layers. Either using additive or subtractive processes. In the case of an additive method the deposition is only done locally where the material is wanted. Classical representatives are printing processes like ink-jet, screen or flexo printing.

In the case of subtractive patterning, a homogeneous layer is first deposited on the substrate and unwanted parts are removed thereafter. These processes are largely used e.g. in silicon technology where complex layer stacks are created by deposition and patterning of metal and oxide or nitride layers. A hybrid approach for the creation of CNN patterns was also presented by several groups where a full layer of SWNTs is first deposited on an intermediate carrier and patterns are then transferred to the final substrate using silicone stamps [68, 89]. This method seems however to be limited to sparse networks. For thicker, well interwoven CNNs the separation of individual patches would be rather difficult and not well defined.

In this work mostly subtractive patterning methods were chosen. In thin-film technology this is done by photolithography after deposition and patterning by wet or dry etching. The closed and stable structure of CNTs makes them chemically inert under most conditions. They can therefore not be solved or chemically etched by solvents or acids. Dry etching methods were therefore chosen for the patterning of deposited CNNs. Once the CNN is deposited, rinsed to remove the surfactant and dried, the patterning procedure can be started. In all cases the sections to be covered by CNN are protected by photoresist patterns. Without further treatment the photoresist is spin-coated onto the CNN, dried and exposed by UV radiation through a photo-lithography mask. The resist patterns are then created using a standard developer followed by rinsing and drying.

5.6.1. CO_2 snow-jet

A vacuum-free method to remove micron and sub-micron sized particles from surfaces is the carbon dioxide (CO_2) snow-jet. A liquid feed of CO_2, typically pressurized at about 50 bar, is filtered and sprayed through a nozzle. The flow can be controlled by a valve (see fig. 5.4).

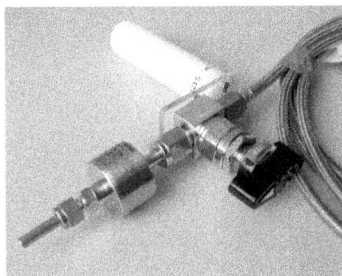

Figure 5.4.: High purity CO_2 snow-jet consisting of nozzle, filter and valve purchased from Applied Surface Technologies.

The rapid expansion of the liquid CO_2, once it passes the small orifice of the nozzle, leads to nucleation of small dry ice crystals - the CO_2 snow. The cleaning effect consists of the momentum transfer of the snow particles in combination with the gaseous jet stream and solving of hydrocarbonic residues by the intermediate liquid phase of the carbon dioxide [121, 122, 126]. Further rapid expansion of the dry snow into gas phase when coming into contact with the surface might additionally help to detach the nanotubes. To avoid icing of the substrate surface that would cover and protect the CNN, the sample can be placed on a hotplate for longer treatments.

This method proved efficient for sparse CNT networks and is quick and easy to apply. For thicker, bundled layers used as TCF, the impact is however not strong enough to efficiently remove the CNN. In addition the more or less localized impact combined with manual handling does not allow for homogeneous treatment of larger surfaces. There are however automated systems on the market. The intensity can further be increased by an additional surrounding flow of compressed air. Such an automated system could serve as an efficient and cost-effective patterning method. In this work the use of reactive ion etching was however often preferred due to it's well controlled and homogeneous impact even on larger surfaces.

5.6.2. Reactive ion etching

Reactive ion etching (RIE) is a dry etching method that is well established in thin-film and silicon technology. A plasma is created by applying a strong electromagnetic RF field of $13.56\,\text{MHz}$ between two parallel plate electrodes (see fig. 5.5). This rips the electrons from the gas molecules which, due to their much higher mobility, oscillate with higher speed between the electrodes as the gas ions. The electrode on which the substrates are placed is insulated for DC currents. Electrons hitting this electrode accumulate a negative charge while the counter electrode and chamber are grounded. In the present research a gas mixture of O_2 and Ar is used. The plasma activates different oxygen radical species which chemically etch the carbon and react to gaseous CO_2 and CO that are evacuated by continuous pumping. The Ar^+ ions are accelerated onto the substrate by the resulting DC field and damage the SWNTs or remove carbon clusters from the SWNTs by momentum transfer. The removed clusters can then further react with the oxygen radicals.

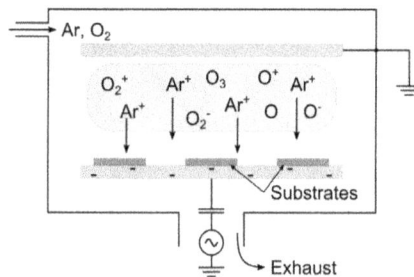

Figure 5.5.: Setup of a reactive ion etching chamber with plasma induced between two parallel plates.

Besides the CNTs the organic photo resist is etched by the RIE at the same time. The much higher thickness of $> 1\,\mu\text{m}$ allows however for several minutes of effective masking before the resist would be completely removed. The high radiation from the plasma plus the ion bombardment leads however to a hardened resist surface which is more difficult to be stripped.

Figure 5.6 shows SEM images of RIE etched patterns with different etch durations. The images show the CNT patterns after stripping of the resist. While electrical measurements revealed a complete loss of conductivity of the CNN after only a few seconds of RIE the SEM images show residual thicker CNT bundles even for prolonged etch durations up to 2 min. An undesired side effect of the longer etch duration becomes visible in figs. 5.6b and 5.6c which are showing clusters forming at the edge of the CNN. Apparently polymer chains are forming under the plasma atmosphere from carbonaceous and organic reagents coming from CNN and photoresist, respectively. These polymers are not removed during resist strip and remain at the edges.

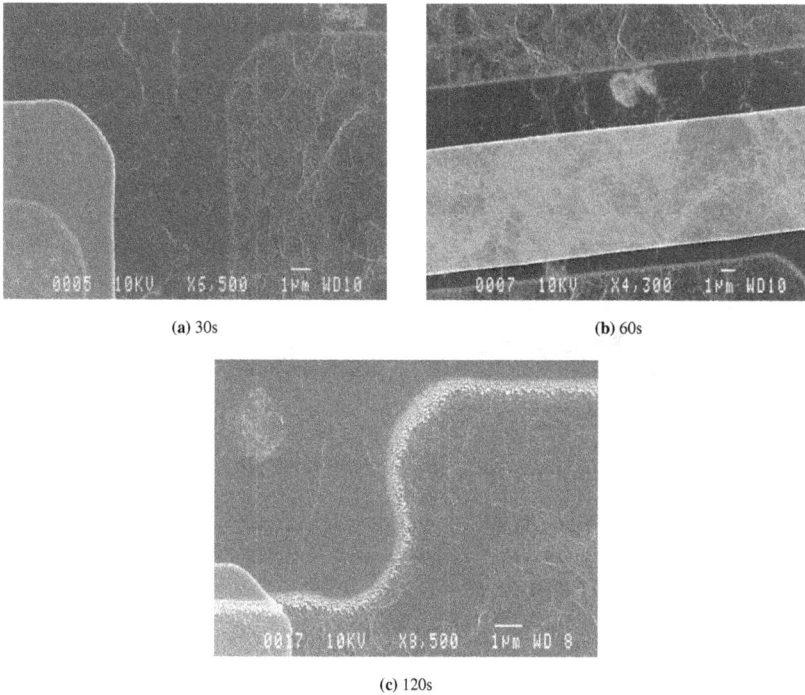

(a) 30s

(b) 60s

(c) 120s

Figure 5.6.: Patterned CNT layers after different RIE etch durations and resist strip; for longer etch durations there rest polymeric residues at the edges after resist strip.

For the thinner spin-coated CNNs in the TFT channels, an etch duration of 30 s is chosen. The thicker, sprayed CNNs that serve as transparent electrode are etched for 1.5 min. For the latter this means that carbon clusters might still be remaining in the etched areas. Electrical tests however never revealed shorts between pixels. The process would still need to be improved for production. This could be done by a) creating CNNs with less or completely without CNT bundles and/or b) an optimized RIE recipe that removes bundles more efficiently and does not form polymers.

5.6.3. Stripping of photoresist

Once the CNN is patterned, the resist used as masking layer needs to be removed. Several solvents for resist stripping were tested [101]. The use of a dedicated photoresist stripper in combination with 10 min ultrasonication turned out to increase the sheet resistance by a factor of 2 or more. Not using any ultrasound treatment during resist strip left however a thin pellicle on the patterned CNN as can be seen in fig. 5.7a. Given that the pellicle is not at the same position as the CNN strips leads to the hypothesis that the outer surface of the resist was hardened or polymerised during RIE etch. By simply soaking the substrates in solvent the inner volume of the resist pattern gets dissolved, the pellicle stays however attached to the pattern edges and is redeposited during rinsing. An efficient resist strip with only minor increase of the

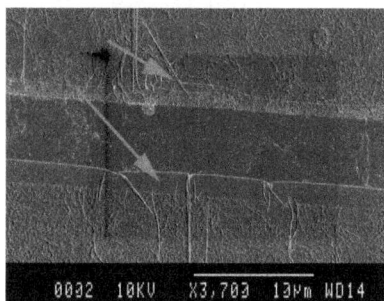

(a) No ultrasound treatment; A thin pellicle of resist is re-deposited (red arrows indicate position).

(b) Sufficient photo resist strip with final recipe.

Figure 5.7.: SEM images of patterned CNN stripes after resist strip.

sheet resistivity was achieved by using 2 subsequent baths of Acetone (see fig. 5.7b). After some minutes of soaking, a short ultrasound treatment removes any residues. A final bath of Isopropanol is used for rinsing as it evaporates slower during N_2 blow-dry. For PES plastic substrates three subsequent baths of Ethanol are used as Acetone attacks the PES core. No remarkable influence of the optimized resist strip process on the optical transmission was detected.

6. Carbon nanotube networks as transparent conductive films

6.1. CNTs vs. other transparent electronic conductors

Indium tin oxide (ITO) is the preferred material for realizing transparent and electrically conductive layers for displays and other applications. It is well established in fabrication processes and has a high electrical conductivity at high optical transparency. It's downsides are high costs of investment for the deposition tools and its brittleness. It also has an elevated refractive index in the order of $\varepsilon_r \approx 1.9$ to 2.0 which causes optical losses as it is not well matched with glass ($\varepsilon_r \approx 1.5$) and air ($\varepsilon_r = 1$) [141]. On plastic and other flexible substrates it breaks easily during flexing or elongation.

There are several alternatives that do not suffer from the above mentioned disadvantages. The organic compound PEDOT:PSS (poly(3,4-ethylenedioxythiophene) polystyrene sulfonate) is a polymer mixture of two ionomers, which has an extended Π-electron complex, similar to CNTs. It can be deposited from a liquid phase by spin-coating or printing techniques. An additive deposition is possible. As this is an organic material it is compatible with flexible plastic substrates. The downside is a lower electrical conductivity for a given transparency and a slightly blueish color of the layer. The performance depends on the pH-value of the material. Ambient influences like humidity and UV radiation are therefore influencing the stability of the product and its longevity.

Silver nanowires are similar to CNTs in a way that these are nanometric wires with a high aspect ratio. They are also deposited from a liquid phase. The Ag nanowires have a diameter in the range of 100 nm. The individual sticks are therefore not flexible and not as easy to deposit with printing techniques as PEDOT:PSS. They have a high electro-optical performance which is close to ITO. A macroscopic network of these wires can still be quite flexible. The Ag nanowires are however prone to oxidation which can deteriorate the conductivity and they therefore need to be protected. The larger diameter compared to SWNTs also creates a higher amount of haze.

The big advantage of carbon nanotubes is their high aspect ratio combined with an extraordinary intra-tube conductivity, mechanical and chemical stability plus a high degree of mechanical flexibility. CNNs can be deposited in a conformal way on 3D surfaces from liquid phase and can withstand high amounts of mechanical stress like bending, stretching or pressure [133]. CNNs were reported to have a refractive index in the range of $\varepsilon_r \approx 1.55$, which is much closer to glass and plastic substrates than in the case of ITO [45]. The following sections show in more detail how carbon nanotubes in the form of thin interwoven networks can be used as transparent conductive film in general and in display applications in particular.

6.2. Morphology and electro-optical performance of spray-coated CNNs

Investigating the different methods listed in chapter 5 for depositing CNTs from liquid phase, spray-coating proved to be the most efficient and most economic for the creation of transparent conductive films. It can easily be applied to large surfaces of glass and plastic substrates and it is able to produce thin and thick layers in a matter of minutes. In an automated production process the spraying could be done even faster. For large surfaces, most of the sprayed material is actually deposited on the substrate. In contrast, spin-coating has only a low ratio of deposited to dispensed material, dip-coating would need a large amount of time without being able to create thicker layers and filtration processes that are often applied in CNN research suffer from scalability and the need for transfer from the filter to the substrate.

Optical microscope images of a CNN layer deposited by spray-coating before and after surfactant removal are given in fig. 6.1. After deposition and drying the surfactant forms flake-like domains. These are the

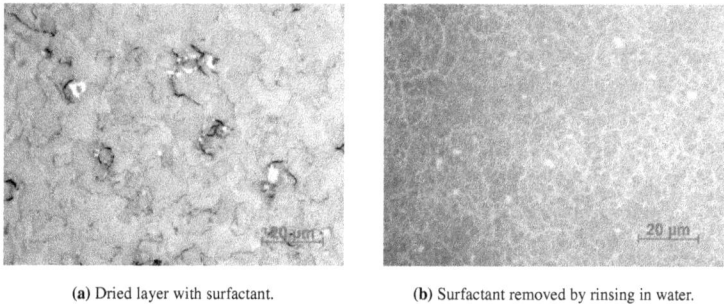

(a) Dried layer with surfactant.   (b) Surfactant removed by rinsing in water.

Figure 6.1.: Spray-coated CNN layer; IsoNanotubes-M suspension, sprayed volume $0.16\,\mathrm{mL/cm^2}$.

individually dried droplets. Once the surfactant is removed, the network of CNTs becomes visible. Thicker bundles are forming, building domains with comparable sizes as seen in the dried surfactant layer. These bundles are likely to form already during drying of each individual droplet as a result of the so-called coffee-stain effect. Due to different evaporation rates at the edge and the center of the droplet, a flow towards the edges transports the suspended particles to the periphery, where they accumulate. During rinsing of the dried layer, the non-polar CNTs adhere strongly to each other and form the bundles. The advantage is that an continuous, interwoven network is created which leads to the high mechanical stability of the CNN. A more detailed view of a CNN is given by the AFM image in fig. 6.2. The topography scan shows height differences in the range of 100 nm for this example.

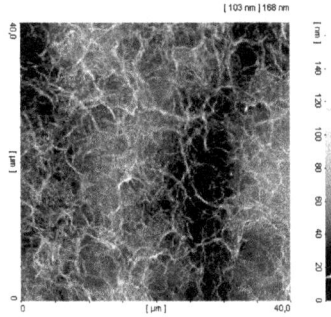

Figure 6.2.: AFM image of a CNN after rinsing.

On the macroscopic level a homogeneous grayish film is observed. Figure 6.3 shows photos of two different materials deposited on glass and plastic substrates after surfactant removal. The suspension volume necessary to achieve a certain transmission depends on the individual suspension as they have different solid content.

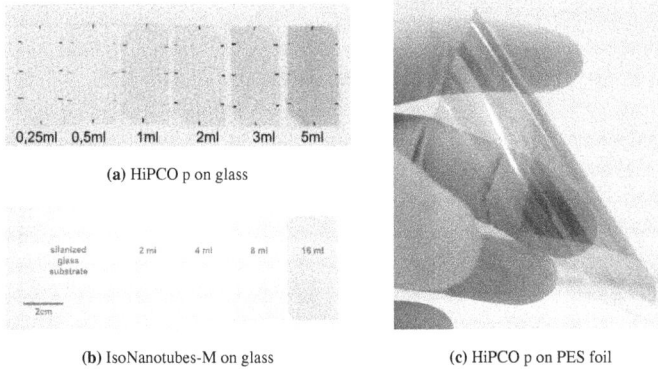

(a) HiPCO p on glass

(b) IsoNanotubes-M on glass

(c) HiPCO p on PES foil

Figure 6.3.: CNT-TCFs of different sources with varied network densities (spray volume noted on individual substrates) after deposition on substrate and rinsing of surfactant.

The goal is to have a high transparency combined with a low sheet resistance. As the bundles absorb more light than the individual nanotubes, it is likely that finer droplets lead to less pronounced bundling and should result in a better electro-optical performance. Different spray parameters like air pressure and distance to substrate were tested to evaluate the influence on the network quality. From optical observations with the microscope shown in fig. 6.4, it can be concluded that the combination of higher air pressure and more

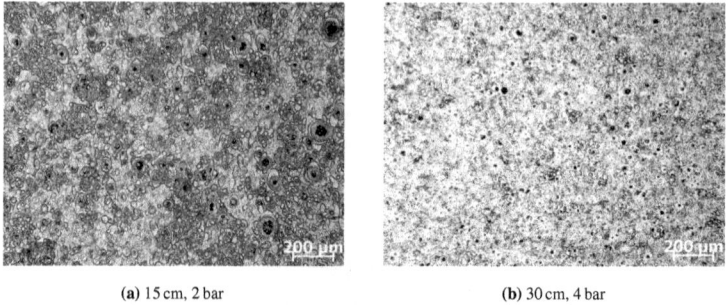

(a) 15 cm, 2 bar   (b) 30 cm, 4 bar

Figure 6.4.: Droplet size depending on distance from nozzle to substrate and airbrush air pressure.

distance leads to smaller domains or droplets. Although a higher pressure and therefore higher gas flow rate is likely to create finer droplets, the influence of the distance between nozzle and substrate could also be related to the density of droplets hitting the substrate. A smaller distance will lead to a higher spray density if all parameters are kept identical. This can lead to intermixing of individual droplets before the water evaporates, resulting in bigger domains. The manual control of the spray process makes it difficult to clearly distinguish between different parameters. Additionally the results are depending on the speed of lateral movement and the skills of the operator. In an industrial spray process these parameters could be better identified and optimized. A finer spray could further improve the electro-optical performance. An ultrasonic vaporizer might create smaller droplets than the classic spray nozzle used in this work. A possible alteration of the CNT suspension would need to be investigated. All results of this thesis are produced with the spray nozzle shown in section 5.2, which is sufficient for a proof of principle.

The optical performance of the produced films is characterized with a transmission spectrometer with a spectral measurement range of 280 nm to 1095 nm. Before each measurement, a baseline is determined with an identical substrate without a deposited CNT film. The measurement data after substraction of the baseline therefore only reflects the influence of the deposited CNN. A sample transmission spectrum in the visible range of 380 nm to 780 nm for each of the used materials is shown in fig. 6.5a. From each CNT type, a measurement was chosen which is closest to $T = 85\%$. The thick lines are smoothed curves, derived from the raw data which is indicated by thin lines. It has to be noted that for short wavelengths below 450 nm the signal is dominated by noise as either the source is too weak in this range or the optical measurement path is attenuating the signal. The form of the transmission curve in this range might therefore be error prone. It can be observed that most spectra have a similar form and are quite flat which results in an almost color-neutral attenuation. In other words the produced TCFs are close to a neutral density filter. Only the IsoNanotubes-M have a more pronounced attenuation for short and long wavelengths. The CIE chromaticity coordinates x and y can be calculated from each spectrum. First the product of the spectrum with the three CIE color matching functions $\bar{x}$, $\bar{y}$, $\bar{z}$ that model the response of a "standardized" human eye is integrated. This results in a set of tristimulus values

$$X = \int T(\lambda)\bar{x}(\lambda)d\lambda, \qquad Y = \int T(\lambda)\bar{y}(\lambda)d\lambda, \qquad Z = \int T(\lambda)\bar{z}(\lambda)d\lambda; \qquad (6.1)$$

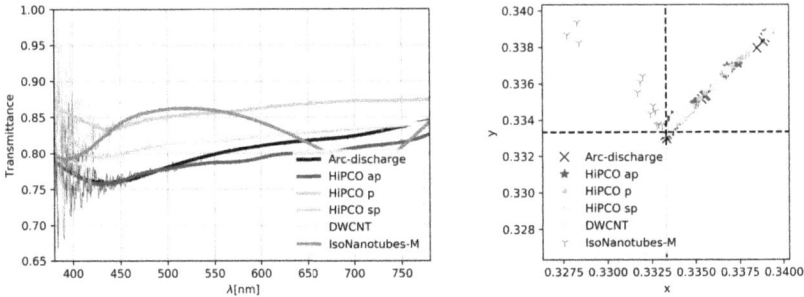

(a) Measured transmittance spectra close to $T = 85\%$; thin lines: raw data, thick lines: smoothed.

(b) CIE chromaticity diagram with x and y coordinates of 108 spectra measured for thicker and thinner CNNs of different feedstock.

Figure 6.5.: Transmission characteristics in the visible spectrum (380 nm to 780 nm) of all CNT materials tested for TCF realization.

the chromaticity coordinates are then obtained by normalization

$$x = \frac{X}{X+Y+Z}, \qquad y = \frac{Y}{X+Y+Z}. \tag{6.2}$$

A Python module from [51] was utilized for the calculation. The results are plotted in fig. 6.5b. Despite the low signal to noise ratio for short wavelengths the CIE chromaticity diagram shows a clear trend. A perfect neutral density filter would have it's coordinates at $(x,y) = (0.\overline{3}, 0.\overline{3})$, independent of it's transmission. The spectra with a high transmittance are close to this point. The thicker the layers become and the more light they absorb, the more pronounced a certain coloration is perceived. The IsoNanotubes-M samples tend towards a cyan complexion (also visible in fig. 6.3b) while all other CNN types rest along the same axis which tends towards yellow. In all cases the chromaticity coordinates stay close to the neutrality point and for a reasonable transmittance of 80 % to 95 %, a coloring is almost not perceivable, which is an advantage over some other TCF materials.

The sheet resistance, measured from the same samples at about the same positions, is combined with the corresponding transmittance at (550 ± 5) nm to characterize the electro-optical performance of the produced TCFs. On each substrate T and $R_\square$ are averaged from three measurement positions to minimize variations caused by possibly inhomogeneous layers. The resulting data points of T vs. $R_\square$, grouped by CNT type, are plotted in fig. 6.6. The behavior can be well modeled to a polynomial curve fit of the form

$$T = -a \cdot R_\square^{-b} + 1. \tag{6.3}$$

The figure of merit φ_{TC} calculated from the polyfit is indicated by dashed lines in the corresponding color on the secondary y-axis. The pair of T vs. $R_\square$ for max(φ_{TC}) is marked as circle on each curve. Table 6.1 summarizes figures of merit and fitting parameters for each material in descending order of performance.

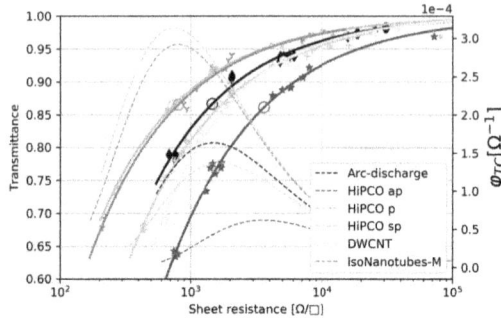

Figure 6.6.: Transmission at (550 ± 5) nm vs. sheet resistance for CNNs from different source material; solid lines are fitted to function $T = -a \cdot R_{\square}^{-b} + 1$; dashed lines corresponding to secondary y-axis are figure of merit φ_{TC} derived from the polyfit with its maximum indicated on each polyfit with a circle.

Table 6.1.: Maximum FOM φ_{TC} in order of descending value with corresponding T and $R_{\square}$ for data shown in fig. 6.6 plus fitting parameters a & b for eq. (6.3).

CNT type	max(φ_{TC}) [1/Ω]	T [%]	$R_{\square}$ [$\Omega/\square$]	a	b
HiPCO p	3.1E-04	86.4	737	9.2	0.64
IsoNanotubes-M	2.9E-04	86.5	804	9.9	0.64
Arc-discharge	1.6E-04	86.6	1460	16.1	0.66
HiPCO sp	1.4E-04	86.3	1675	14.4	0.63
DWCNT	1.2E-04	87.8	2263	32.0	0.72
HiPCO ap	6.2E-05	86.1	3629	21.3	0.61

As expected the maximum of φ_{TC} lies at a transmittance of 86 % to 88 % for all products. The best performing HiPCO p SWNTs achieve 737 $\Omega/\square$, while the least performing HiPCO ap SWNTs achieve this transmittance at 3.6 k$\Omega/\square$. This is likely caused by the high amount of impurities that the ap grade contains (compare table 4.1). The super purified HiPCO grade however achieves only average performance, which is not logical at first glance. A possible explanation is that the sp grade powder contains large bundles that are much harder to exfoliate. Another possible explanation is that the more severe purification process of the sp grade might have damaged or shortened the SWNTs. Arc-discharge and DWNTs perform similar to the HiPCO sp grade. The highly enriched metallic IsoNanotubes-M behave almost identical to the HiPCO p SWNTs. As already explained in section 2.1, a purely metallic CNN does not necessarily lead to a higher conductivity. The s-SWNTs are doped from the ambient atmosphere, which leads to lower Schottky barriers and higher density of states in the s-SWNTs. The IsoNanotubes-M were purchased as a compressed pellet which was much more difficult to disperse. It is therefore also possible that the performance is limited by more pronounced bundling compared to the p grade HiPCO CNTs.

Table 6.2 puts the so far discussed results in context with CNT-TCFs produced by other research groups. Dry deposition methods have the advantage of not having to exfoliate the bundles and avoid to pass by a wet phase that might add further impurities or damage the CNTs. CNT-TCFs produced by such methods

Table 6.2.: Record values from literature for CNT-TCFs produces with different dry and wet deposition techniques.

Citation	CNT type	Deposition Type	Method	Dispersion	$R_\square$ [$\Omega/\square$]	T	φ_{TC} [Ω^{-1}]	Dopant	Substrate
[3]	Canatu CNB	Dry	Direct Dry Printing	-	100	0.94	5.4E-03	?	
[106]	Floating catalyst	Dry	Dry Transfer from Filter	-	224	0.90	1.6E-03	none	Quartz
[23]	Floating catalyst	Dry	Transfer	-	417	0.89	7.3E-04	none	
[106]	Floating catalyst	Dry	Dry Transfer from Filter	-	86	0.90	4.1E-03	AuCl3	Quartz
[69]	Floating catalyst	Dry	Transfer	-	110	0.90	3.2E-03	HNO3	
[44]	CVD	Wet	Spray-coating	Surf. aq.	650	0.86	3.4E-04	none	PC
This work	HiPCO	Wet	Spray-coating	Surf. aq.	737	0.86	3.1E-04	none	PES
[93]	Arc-discharge	Wet	Vacuum filtration + transfer	Surf. aq.	110	0.78	7.6E-04	none	PET
[90]	HiPCO	Wet	Dip coating	Acid	100	0.90	3.5E-03	H2SO4	
[43]	CVD	Wet	Vacuum filtration + transfer	Acid	60	0.91	6.4E-03	HSO3Cl	PET
[96]	CVD	Wet	Spray-coating	Ammonia	311	0.93	1.6E-03	HNO3	
[48]		Wet	Spray-coating	Solvent	440	0.87	5.6E-04	MoOx	Glass@220 °C
[48]		Wet	Spray-coating	Solvent	100	0.85	2.0E-03	MoOx + 450 °C	Glass@220 °C
[137]	Laser ablation	Wet	Ultrasonic spray	Solvent	167	0.77	4.5E-04	HNO3	Glass@220 °C
[82]	CVD	Wet	Spray-coating	Surf. aq.	240	0.90	1.5E-03	HNO3	PET
[32]	Arc-discharge	Wet	Spray-coating	Surf. aq.	86	0.80	1.2E-03	HNO3/SOCl2	PET
[93]	Arc-discharge	Wet	Vacuum filtration + transfer	Surf. aq.	37	0.76	1.7E-03	HNO3	PET

typically show a higher φ_{TC}. In many cases, the direct deposition on the substrate is however not possible due to process temperatures. Independent of the deposition technique, the electro-optical performance can be increased significantly by doping the final layer. However, it should be noted, that these doping methods are often not long-term stable.

In this work the main focus was put on the display application. A stable performance was therefore preferred instead of record values. Wet processed TCFs from other research groups without further doping are in the same order of magnitude as the one presented here. The developed processes for dispersing and deposition therefore proved to be efficient although there is likely to be room of improvement. An important prerequisite for optimizations are more sophisticated characterization methods for suspensions and fabricated films, though.

6.3. Contacting CNNs with metal layers

In many applications the TCF layer is electrically connected by metal contacts. Specifically, this is the case in AMLCDs, where the narrow buslines are realized by metals such as AlNd, Cr, Mo or Ta [112]. In contact areas for flexible PCB bonding or point probes, a pad of a thin metal is also more robust than the CNN itself. A low contact resistance between the CNN and the used metal is therefore very important. A screening of several metals was realized with not only the contact resistance as criterion but also usability on flexible plastic substrates. Loosing the compatibility to flexible and especially plastic substrates due to the choice of metal buslines would exclude one of the most advantageous application fields of CNNs.

6.3.1. CNN-metal contact resistance

The choice of metals is based on established materials for active-matrix LCD and OLED applications at the Institute for Large Area Microelectronics. Additionally high work function metals known from literature to give good performance in CNT-TFTs are also investigated. The latter ones are most often evaporated onto the CNT layer and patterned by lift-off. In AMLCD production the metal is usually deposited by sputtering and patterned before the deposition of the TCF. In the sputter process, metal clusters hit the substrate surface at high velocity. The impact energy can damage the atomic structure of the nanotubes, which would compromise the electrical characteristics of the CNN. Metals deposited by sputtering are therefore only tested below the CNN.

Using the test structures described in section 2.5.2, the contact between spray coated CNN with a sheet resistance of $2\,\mathrm{k\Omega}/\square$ to $4\,\mathrm{k\Omega}/\square$ and different metals was determined. This was done for the top contact configuration where the metal is evaporated onto the CNN and patterned by lift-off (table 6.3a) as well as the bottom contact configuration where the CNN is sprayed on pre-formed metal patterns (table 6.3b). As the contact resistance R_c is scaling with the interface area A, the specific contact resistance

$$\rho_c = R_c \cdot A \qquad [\mathrm{m\Omega cm^2}] \qquad (6.4)$$

is calculated as an area-independent measure.

Material	deposition method	specific contact resistance $\rho_c\ [\mathrm{m\Omega cm^2}]$	Work function of metal [eV]
Pd	evaporation	35	~ 5.4
Ti	evaporation	208	4.3

(a) Top contact configuration; metal deposited on top of CNN and patterned by lift-off.

Material	deposition method	specific contact resistance $\rho_c\ [\mathrm{m\Omega cm^2}]$	Work function of metal [eV]
AlNd	sputtering	(>20000)	~ 4.2
Au	sputtering	75	~ 5.3
MoTa	sputtering	90	~ 4.7
Pd	evaporation	60	~ 5.4
Ti	evaporation	790	4.3

(b) Bottom contact configuration; CNN deposited on patterned metal.

Table 6.3.: Contact resistance between different metals and CNNs deposited by spray-coating.

The evaporated metals Pd and Ti could be tested in both, top and bottom configuration. In direct comparison a top contact gives a significantly lower contact resistance. In the case of Pd the bottom contact resistance is almost twice as high, for Ti it is almost 4 times higher. This is due to several facts.

- The 2.5 dimensional CNN is better covered and voids are filled leading to a larger effective contact area

- Conduction can be impeded by native oxides forming on bottom contacts made of Ti before the deposition of the CNN
- Residual surfactant molecules at the bottom contact - CNN interface might be harder to remove

The very high contact resistance in the case of AlNd can be explained by a native oxide forming on the metal surface. Individual measurements showed a high spread, but were always very high, thus forbidding practical use. The other bottom contact metals do not reach the very low resistance of top contact Pd but stay in the same order of magnitude. For the tested metals, the contact improves with rising metal work function. However also the moderate work function of MoTa leads to a reasonably low specific contact resistance of $90\,m\Omega cm^2$.

The determined values are in the same order of reported specific contact resistance to Ag (work function of $\approx 4.5\,eV$) of $91\,m\Omega cm^2$ for undoped and $20\,m\Omega cm^2$ for p-doped CNNs with $350\,\Omega/\square$ and $200\,\Omega/\square$, respectively [60]. Values for metal-ITO contacts are almost one order of magnitude smaller. The elevated specific contact resistance in the case of CNNs can be a serious restriction for more demanding applications and might be caused by Schottky barriers to the large amount of semiconducting nanotubes, the non-planar contact of CNN and metal, possible chemical reactions at the interface and residues like surfactant [60].

From the tested metals Au, MoTa and Pd lead to a sufficiently good electrical contact for the application in voltage driven AMLCDs, where the liquid crystal layer and the thin-film transistor add significantly higher resistance in the electrical circuit.

6.3.2. Metal layers compatible with flexible plastic substrates

The experimental samples in the preceding section were realized on glass substrates. A major advantage of CNT-CNNs over classic oxide TCFs is the bendability and compatibility with flexible plastic substrates. It is therefore of interest to use metal buslines that are compatible with these substrate materials. Some of the above mentioned metals where deposited on PES foils that already have a thin Ta_2O_5 adhesion layer. It is important to keep the thickness of this adhesion layer as thin as possible, because a thick layer would be brittle. A thickness of 30 nm proved to be enough as adhesion promoter and etch-stop layer. At the same time no cracking could be diagnosed after bending of the substrates. Evaporated Pd and Ti were not included in the test as their deposition is very time consuming. The fast sputter deposition was favored. Another advantage of this process is a better adhesion between layer and subtrate due to higher impact of the deposited atoms and clusters. Table 6.4 lists the tested metals. Unfortunately, MoTa, which was chosen as contact metal in glass based AMLCDs, is not compatible with the plastic substrate. After patterning, cracks were observed right away (see fig. 6.7a). The same holds for Au on Cr. The later is used as adhesion promoter due to often poor adhesion of the gold layer on organic and inorganic surfaces. The brittle Cr also quickly formed cracks (see fig. 6.7b). It was also observed that Au peeled of from the Cr layer. The etch-stop layer of Ta_2O_5 can also serve as adhesion promoter, allowing Au layers to be deposited directly on it. The adhesion proved to be sufficient to withstand a scotch tape test. Both, Au and AlNd did not show

Table 6.4.: Qualitative summary of metal layer usability for contacting CNNs on PES foils with Ta_2O_5 layer.

material	configuration	adhesion	contact resistance	ductility
AlNd	bottom contact	+	- -	-
Au	bottom contact	+	+	++
Cr/Au	bottom contact	-	+	- -
MoTa	bottom contact	+	+	- -

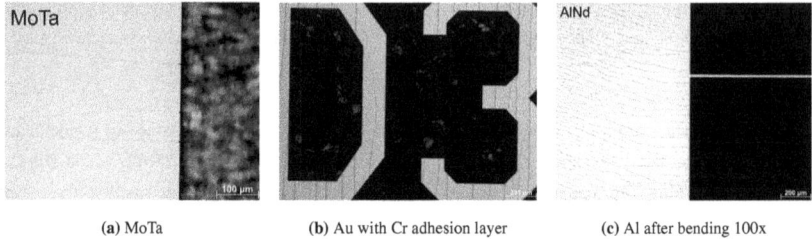

(a) MoTa   (b) Au with Cr adhesion layer   (c) Al after bending 100x

Figure 6.7.: Cracks in metal layers on PES foil with 30 nm Ta_2O_5 adhesion layer.

any cracks after deposition and patterning. After a more severe bending test with 100 repetitions, the AlNd showed many cracks despite it's high ductility (see fig. 6.7c). Nevertheless, the contact between AlNd and CNTs is not sufficient due to the aluminum's natural oxide layer.

A 50 nm Au layer on 30 nm Ta_2O_5 turned out to be the best candidate. No cracking or delamination was observed, even after the bending test. Additionally, it has the second smallest contact resistance of the metals tested in bottom contact configuration.

6.4. Capping layer and liquid crystal orientation layer

The first polymer dispersed liquid crystal displays (sections 2.4.2 and 6.5.1) were realized without further treatment of the CNN electrodes. Many of the display segments showed short-circuits between the front and back electrodes. An analysis of the CNN surface did not show any features that could have breached the 14 μm cell gap. The cause turned out to be the spacer balls used for assuring the cell gap of the displays. These are sprayed on one of the substrates. During assembly of the display sandwich even small movements between the two plates lead to a displacement of the spacer balls. Figure 6.8 clearly shows how the spacer ball pics up a trace of CNN during such a movement. The result is a conductive path between the opposite electrodes.

To increase the mechanical robustness of the CNNs and to avoid such shorts, a capping layer is introduced. In anticipation of the realization of twisted nematic LCDs, a polyimide named SUNEVER® SE-130 from Nissan Chemicals Industries Ltd. used as orientation layer in such displays was chosen as capping layer.

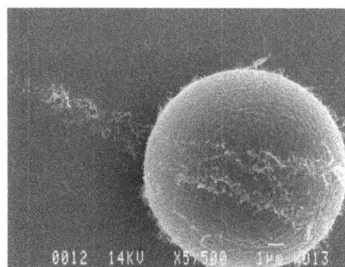

Figure 6.8.: SEM image of a spacer ball with a trace of CNN picked up by rolling on a CNN electrode.

This assures the compatibility with the LC mixture and allows simultaneous use as capping and orientation layer. The polyimide layer is produced by spin-coating and oven-curing for imidization (paramters in appendix A.2). A range of possible layer thicknesses was determined by spin-coating different dilutions at a set of revolution speeds. Layers of a nominal thickness are then deposited on CNNs using the parameter set as evaluated on bare substrates.

Figure 6.9 shows SEM images of CNT layers with capping layers of nominally 20 nm to 200 nm thickness. At 20 nm the bigger CNT bundles are still visible. At 70 nm the CNN is already well smoothed. With a capping layer of 200 nm nominal thickness no surface roughness is visible any more in the SEM image and a particle had to be focused to make sure that the surface is visible.

A better quantification of the surface roughness is possible by using atomic force microscopy (AFM). Scans of 10 μm × 10 μm are shown in fig. 6.10. Roughness values taken from these areas are listed in table 6.5. The AFM scans show clearly that even at 200nm thickness the PI layer still shows some waviness. Obviously this is strongly dependent on the scanned location as can be seen in the SEM image in fig. 6.9a.

Table 6.5.: Roughness of PI-CNN layers for different PI thickness; all values in nm; S_y: peak to peak profile height, S_z: ten point height, S_a: average roughness; see appendix B for exact formula of roughness values.

Nominal PI thickness	S_y	S_z	S_a
20	115	91	6.39
70	114	101	6.49
130	48.8	45.4	3.67
200	22.9	21.5	1.62

In contrast to the optical appearance in the SEM images, the roughness values for the samples with 20 nm and 70 nm PI layer are not very different. This holds for both - the overall profile height as well as the average roughness. In this particular experiment, the 70 nm sample even had higher roughness values which can be explained with the inhomogeneities due to bundle formation. A layer thickness above 100 nm is required for planarization. At this thickness the voids between the bigger bundles are getting filled. At

(a) 20 nm

(b) 70 nm

(c) 200 nm, the particle is only used to demonstrate
the focus of the otherwise featureless image.

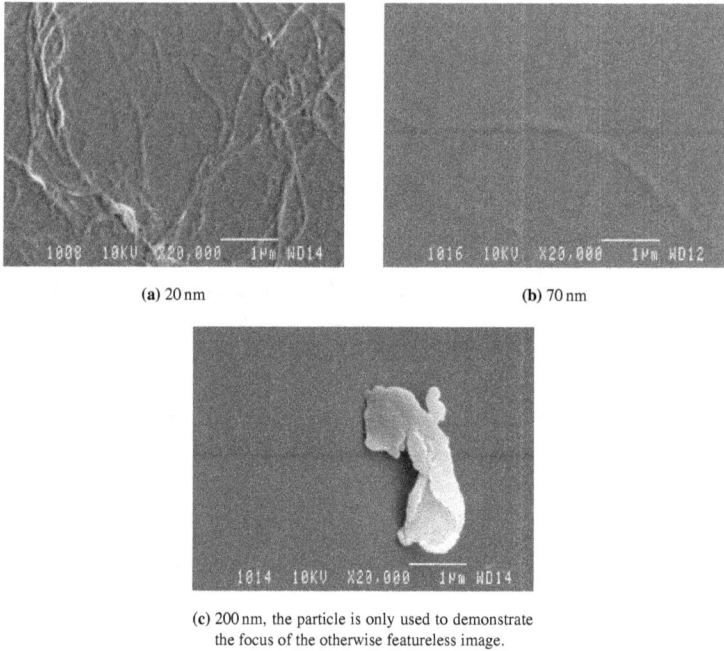

Figure 6.9.: SEM images of polyimid layers of different thickness on CNT networks.

200 nm the average rougness of $S_a = 1.62$ nm is getting close to a glass substrate average roughness of 0.7 nm to 1 nm.

The spin-coated PI layer therefore seems to soak into the CNN rather than coating it conformingly. This impression is further backed up by scotch tape tests that show an increased adhesion of CNNs to the substrate when using capping layers. This holds already for very thin PI coatings, indicating that the PI improves the adhesion at the interface between substrate and CNN.

6.5. Realization of display demonstrators

The main focus of this work is the use of carbon nanotubes in display applications. Although the so far presented recipes and analyses are important building blocks to achieve this goal, the realization of actual LCD demonstrators gives a much more direct feedback how well this goal can be achieved.

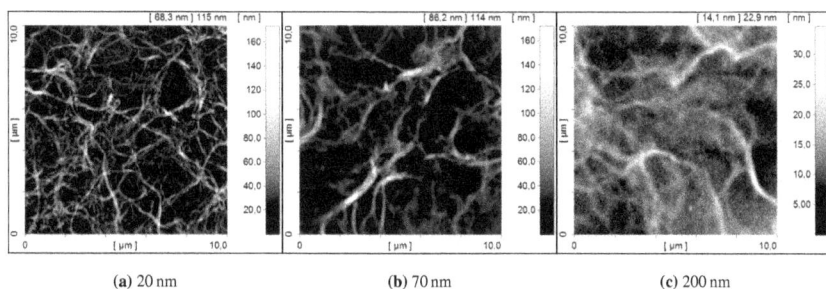

(a) 20 nm   (b) 70 nm   (c) 200 nm

Figure 6.10.: AFM images of polyimid layers of different thickness on CNT networks; note the differences in height scale.

6.5.1. Polymer dispersed LCDs

The setup and functioning of a PDLC is explained in section 2.4.2. It was chosen as demonstration vehicle because its realization does not demand any further treatments after deposition and patterning of the TCF. Once the TCF electrodes are defined on both substrates, spacer balls are sprayed on one substrate and a glue frame is applied to the other one. The glue frame is cured by UV exposure once the sandwich is assembled and aligned. The void in the sandwich is filled with a mixture of pre-polymer and liquid crystal. This is done by capillary force. The filling process can be accelerated by applying a slight vacuum at the other end of the cell. Once filled, the mixture is cured by UV irradiation. This causes the pre-polymer to polymerize which leads to a so-called polymerization induced phase separation (PIPS). By adjusting the energy density of the UV radiation, the size of the LC spheres in the polymere matrix can be controlled [29]. Detailed process parameters for the fabrication of PDLCDs are listed in appendix A.3.

Initial demonstrations used symbols and buslines of large dimensions to make sure no open circuits appear due to voids in the CNN. It could however quickly be demonstrated that higher resolution patterns (see fig. 6.11a) and narrow buslines (see fig. 6.11b) down to 5 μm width with aspect ratios (busline length/width) of 10'000:1 and more can repeatedly be realized. Initial demonstrators often showed shorted segments between electrodes and counter-electrodes. As explained in section 6.4 this problem was attributed to spacer balls that take up traces of the CNN when moved between the substrates during the assembly process. This can be completely avoided by application of a capping layer.

First flexible plastic displays with CNT-TCFs were also realized as PDLCDs. Figures 6.11c and 6.11d show an example in flat and bent configuration. The contrast of the plastic based display is strongly reduced in comparison to the glass based display, even in the flat configuration. This is due to a less well defined gap between the substrates. For the glass version, it is sufficient to apply spacers on the complete surface and fix the alignment with two lines of glue at opposed edges (as visible in figs. 6.11a and 6.11b) to achieve a constant cell gap before, during and after filling. The plastic substrates are however already slightly bent. As there are no external constraints, the cell gap can be bigger than designed. During filling, the spacers might

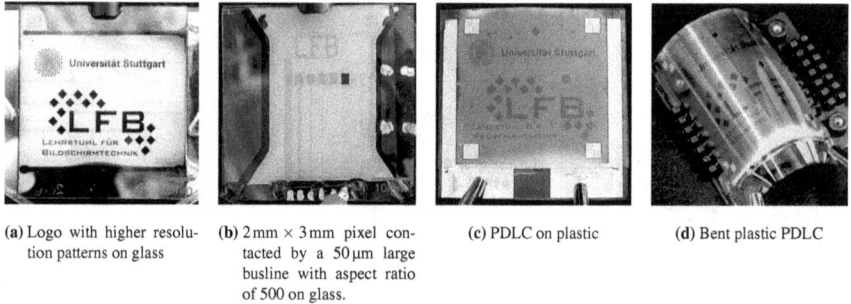

(a) Logo with higher resolu-   (b) 2 mm × 3 mm pixel con-   (c) PDLC on plastic   (d) Bent plastic PDLC
tion patterns on glass   tacted by a 50 µm large
                                busline with aspect ratio
                                of 500 on glass.

Figure 6.11.: Examples of polymer dispersed LCDs on glass and plastic substrate with transparent SWNT
electrodes; all displays have the same size.

however also be washed away with the filling front as the spacers are not kept in place by both substrates. The
white non-addressed PDLC matrix gets even more transparent when bent. Additionally bending might create
electrostatic charges that can align the LC droplets. A second possibility are mechanical stress induced
optical changes. During bending the inner substrate has to be compressed while the external one needs to
be elongated to keep the same cell gap. As a result there is a big force trying to compress the PDLC layer.
This compression is further accentuated if the spacer density already got reduced during filling.

Another aspect is the fact that the patterned TCF is clearly visible in the high contrast glass PDLCDs.
During fabrication the aim was to achieve a sheet resistance below $1000 \, \Omega/\square$ which results in about 85 %
transmittance. Superposed on the white PDLC layer, the structures become clearly visible. The thickness of
the CNN might be further decreased to have a higher transmittance. For larger segment displays with longer
buslines, this can lead to excessive series resistance, which might be compensated by increased addressing
voltages at the cost of power consumption. Another workaround is presented in the next section.

Despite the mentioned flaws the produced displays were the first realization of segmented LCDs containing CNT-TCFs in the world and demonstrate the suitability of the material itself as well as the developed
processes.

6.5.2. Twisted nematic directly addressed displays

PDLCDs only found their way into some niche products because of the aforementioned drawbacks. Today's
LCD products almost exclusively use polarization effects instead of scattering. For several decades, twisted-
nematic LCDs were dominating the display marked from simple seven segment displays over telephones and
PC monitors up to large scale televisions. In recent years the TN technology got more and more replaced
by higher quality products using other liquid crystal modes like in-plane switching and vertical alignment,
which offer better contrast, viewing angle and switching speeds. TN-LCDs nevertheless are still widely

used these days and it is the chosen mode for the realization of the following demonstrators. The results demonstrated here are also relevant to the more advanced LC modes mentioned above.

The two main differences between TN-LCD and PDLCD, are the need for LC orientation layer and polarizing filters (as described in section 2.4). The polarization filters are simply laminated onto the outer surfaces of the display at the end of the fabrication process. The LC orientation layer however is in direct contact with the CNT-TCFs. It is therefore a crucial development. The material of the capping layer presented in section 6.4 was already chosen with the LC orientation in mind. This polyimide needs to be mechanically rubbed so that the LC molecules align in the direction of rubbing. In the case of thin-film TCFs like ITO a thin PI layer in the range of some 10 nm thickness fully covers electrodes or substrate where there is no TCF with good adhesion. It was shown in section 6.4 that the PI layer soaks into the CNN with the network topography still visible. A very thick layer could completely cover the CNN. Excessive PI thickness would however absorb a part of the electrical field strength, leading to increased addressing voltages.

The standard rubbing process is using a rotating, velvet coated cylinder. The substrate is passing below this cylinder with a defined rubbing pressure and speed. As the CNT-TCFs are not very scratch resistant, this rubbing machine was initially not used to avoid potential contamination of the velvet cloth. A soft carbon fiber brush was used instead to create the necessary alignment [101]. The brush is brought into contact with the substrate and further approached by 1 mm. With 30 passes of the brush at 150 mm/s a comparable optical result as with the standard tool could be achieved on ITO test cells. No mechanical damage to the PI nor the CNN could be stated. The alignment works even with PI layers that are nominally thinner than the CNN topography. A standard thickness of 60 nm was chosen. Detailed process parameters for the fabrication of TN-LCDs are listed in appendix A.4.

To render the segments and supply lines less visible a dummy pattern was introduced that covers the areas where the TCF would normally be completely removed (see fig. 6.12a). The hexagonal pattern has 10 μm

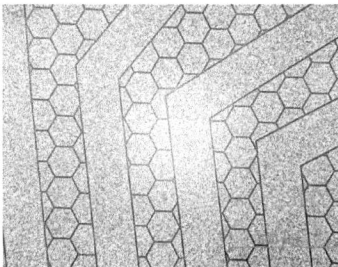

(a) Hexagonal dummy pattern between segments and supply lines.

(b) 7 segment style clock; brighter squares are alignment patterns with CNN only on one substrate.

Figure 6.12.: TN-LCDs with CNT-TCFs

spacing between hexagons and electrodes. No short circuits between individual segments were detected even though residues of the CNN were visible in SEM images. This underlines the efficiency of the CNN patterning process.

A well working TN-LCD demonstrator on glass is shown in fig. 6.12b. Thanks to the dummy pattern no individual segments or supply lines are visible, although at the cost of an overall darker display. A difference is only visible in the 4 square-shaped alignment patterns, where the TCF is only present on one substrate. Such alignment patterns would usually not be placed in the visible area. Contact areas outside the LC cell were reinforced by Ti evaporated through a shadow mask. A bottom contact metal is also possible and will be demonstrated below.

Iso-contrast measurements were performed on a TN display with CNT-TCF and on a ITO control cell. The diagrams are shown in fig. 6.13. The ITO cell has a higher maximum contrast of 1:187. The CNT cell only

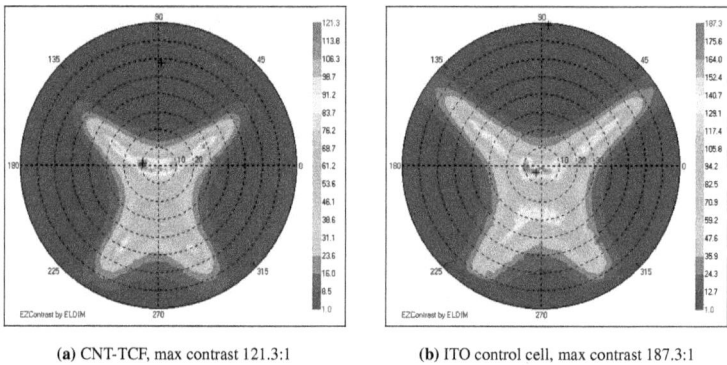

(a) CNT-TCF, max contrast 121.3:1   (b) ITO control cell, max contrast 187.3:1

Figure 6.13.: Iso-contrast measurements of TN cells at $V_{LCD} = 5$ V.

reaches 1:121. This could be caused by a higher series resistance of buslines resulting in a lower LC-voltage. Another difference is the lower viewing angle at 2 and 10 o'clock. Two possible causes might be a) a not as good alignment in the off-state and/or b) the topography of the CNN. It should be noted that both effects are strongly dependent on the orientation of the laminated polarizers and the final cell gap. Both can have some variation given the manual processes. Despite the darker appearance, the realized CNT TN-LCDs show quite similar behavior compared to ITO samples.

The main advantage of the CNT-TCF, its flexibility, is demonstrated by realizing flexible TN-LCDs. A weak point of the plastic PDLCDs were the spacers that are not well enough fixed. For TN-LCDs this problem is increased because even if the vacuum filling process might not sweep away the spacers it turned out that they are transported towards the sealing frame during repeated bending. Adhesive spacers were therefore introduced [115]. These have a coating which adheres to the substrate after an annealing step. With all these modifications in place, plus the ductile gold layer introduced in section 6.3.2, flexible TN-LCDs with CNT-TCFs could be realized. Figure 6.14 shows two demonstrators in a bent state. Even during bending the display continues to work and no broken lines were observed. The standard polarizing films render the display rather stiff. For a one time conformable display lamination to a curved surface is possible. For a real flexible display thinner or less stiff polarizing films should be applied.

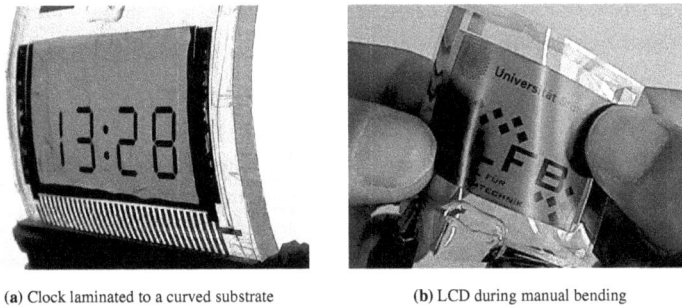

(a) Clock laminated to a curved substrate   (b) LCD during manual bending

Figure 6.14.: Flexible TN-LCDs with carbon nanotube TCF.

6.5.3. Active-matrix LCD

The so far presented displays are all directly addressed, segmented displays. These are sufficient for displaying simple information. In contrast, modern static and mobile devices from smartphones to televisions nowadays use active-matrix displays (compare section 2.4.4). An important question of this work is whether CNT-TCFs can also be applied to AMLCDs. A full color quarter-VGA resolution display (320xRGBx240) was realized where the pixel elektrodes on both backplane and frontplane are realized with carbon nanotube networks. The results are based on a classic a-Si:H back channel etch process established before this work and an poly-Si AMLCD design that was adapted to the a-Si:H process where necessary. The main characteristics of the design are listed in table 6.6. Some test patterns were included for the in-situ characterization of the CNT-TCF (see section 2.5.2). It has to be underlined that the use of CNT-TCFs does not imply any specific design or production changes. A standard AMLCD process can be easily adapted by simply exchanging the process block of TCF deposition and patterning.

Table 6.6.: Main characteristics of the AMLCD design.

LC mode	Twisted Nematic
Display active area	80 mm x 60 mm - 4 inch diagonal
Resolution	320 x RGB x 240 pixel
Sub-pixel size	83 μm x 250 μm - 102 ppi
TFT technology	a-Si:H
TFT channel length	10 μm
TFT channel width	50 μm
Pixel aperture ratio	52%

6.5.3.1. AMLCD process

This section describes the process flow used for the production of the AMLCD demonstrator. Detailed process parameters are listed in appendix A.5. First the backplane including the active matrix and the frontplane with the color filters are processed in parallel. Then both substrates are assembled and the resulting cell is filled with liquid crystal mixture.

Backplane process   The active matrix on the backplane is realized in a bottom-gate, top-contact or inversed-staggered topography by a standard back-channel-etch (BCE) process flow. Starting from the bare glass substrate the first metalization layer is deposited by sputtering MoTa and patterning by photolithography and wet etching with the first mask. In the next step the complete stack of Silicon nitride (Si_3N_4), amorphous Silicon (a-Si) and n-doped a-Si (n^+a-Si) is deposited by plasma-enhanced chemical vapor deposition (PECVD) without breaking the vacuum in between. The three layers serve as gate dielectric, semiconductor and injection layer between semiconductor and metalization layer to prevent the formation of Schottky contacts. The source/drain metalization of MoTa is deposited directly afterwards by sputtering (see fig. 6.15a).

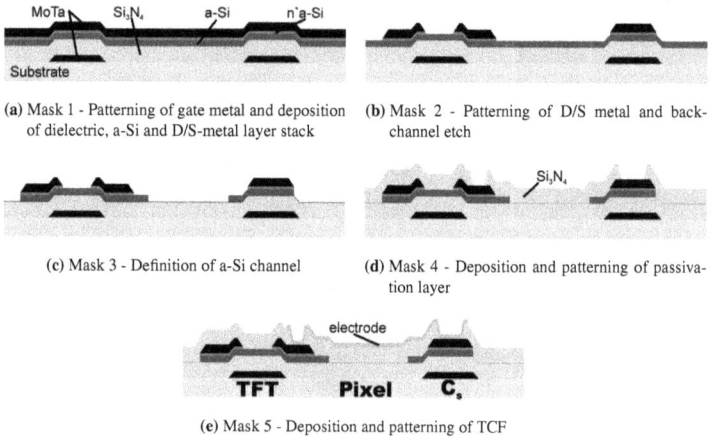

(a) Mask 1 - Patterning of gate metal and deposition of dielectric, a-Si and D/S-metal layer stack

(b) Mask 2 - Patterning of D/S metal and back-channel etch

(c) Mask 3 - Definition of a-Si channel

(d) Mask 4 - Deposition and patterning of passivation layer

(e) Mask 5 - Deposition and patterning of TCF

Figure 6.15.: AMLCD backplane process flow

Source buslines, storage capacitor top electrodes and source/drain contacts are defined with the second photo-mask and wet chemical etching. This metal pattern serves as self-aligned mask for the following back-channel-etch. Here the n^+a-Si layer is removed from all areas that are not covered by the source/drain metal with a plasma etch (see fig. 6.15b). This step is crucial for the later performance of the TFTs. To avoid elevated leakage currents, the n^+a-Si needs to be completely removed while not etching too deep into the a-Si channel.

The semiconductor channel islands are defined with the 3rd mask (see fig. 6.15c). Like the n^+ layer, the a-Si is also masked by the source/drain metalization. The resist pattern further masks the TFT channels, busline crossings to avoid problems with undercuts as well as narrow regions at the side of drain-contact and storage capacitor top electrode. The latter two are a precaution to improve step coverage of the subsequently deposited pixel electrode layer. This precaution comes from design rules for ITO electrodes. It is probably not necessary for the carbon nanotube TCFs as the CNN covers step heights quite well [98].

The TFTs are then passivated by a second Si_3N_4 layer. This avoids contamination of the sensitive back channel that would result in degradation of the TFT characteristics. Vias to the TFTs pixel contact and storage capacitors are defined with the 4th mask and a plasma step (see fig. 6.15d). This etch step also creates vias to the gate metalization by etching through the dielectric layer. The selective etch process stops at the metal electrodes and allows therefore to realize two different etch depths in one step.

The backplanes are finalized by applying the self-assembled monolayer and subsequent deposition of the carbon nanotube suspension via spray coating which serves as pixel electrode. The layer is patterned by the 5th photolithographic mask combined with an RIE etch (see fig. 6.15e). This last subfigure gives the final cross section of the functional backplane parts. TFT, pixel electrode and storage capacitor C_s are indicated. The lateral design of a single pixel is given in fig. 6.16.

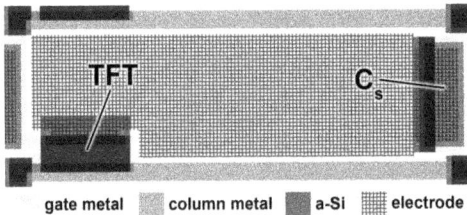

Figure 6.16.: Lateral design of a single pixel. The positions of thin-film transistor (TFT) and storage capacitor (C_s) are indicated.

Frontplane process While the backplane contains the electrical active parts of the display the frontplane holds the color filter that is used to create a full color image. The process flow also starts from the bare glass substrate. The first layer is the so-called black matrix which shields the optically non-active regions. It yields to a higher contrast ratio of the display by blocking all light that is not controlled by the pixels and color filters. The effect is further increased by using a black material that also absorbs light coming from the viewers side. It is defined by sputter deposition of a metal oxide / metal stack that is optimized to give a black appearance and is defined by a photolithographic step and wet etching (see fig. 6.17a). The color filters are created from photosensitive color resists. The process flow is deposition by spin-coating and hotplate drying, exposure through a mask and development. The resists are photonegative materials and are further cured by UV exposure and baking. All three colors are realized subsequently. The additional curing step is therefore necessary to avoid chemical cross-reactions during the processing of the different resists (see fig. 6.17b). Before the transparent conductive film can be applied a planarization layer is deposited on the color filters. This is done by spin-coating a resist and curing. The counter electrode is defined by deposition of the carbon nanotube TCF. This layer is not patterned to the pixel level (see fig. 6.17c). The frontplanes are then cut to size.

(a) Black matrix on glass substrate

(b) Red, green and blue color filters

(c) Planarization layer and transparent conductive film

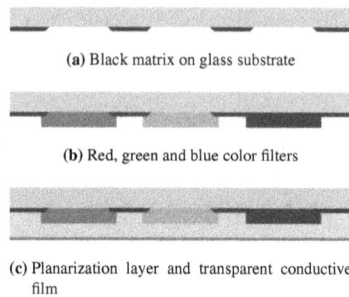

Figure 6.17.: AMLCD frontplane process flow

LCD process   Once both front- and backplanes are finalized the LCD modules can be built. First the Polyimide is spin-coated on both substrates. Areas that need to be electrically contacted are cleaned with a solvent soaked q-tip. After curing, the LC orientation direction is created by rubbing the PI with the standard machine using a rotating velvet cylinder as it turned out that the CNT/PI layer is also not damaged by the standard process. The glue frame is dispensed on the backplane, while spherical spacers are sprayed onto the frontplane. The contact between backplane and frontplane is established by silver glue in all corners. Both substrates are then put together, aligned and cured in a mask aligner. The resulting cells are filled with liquid crystal mixture by vacuum filling. The AMLCDs are finalized by laminating crossed polarizing filters to front- and backplane and bonding the gate and source drivers.

6.5.3.2. Results

With all the developed processes presented so far, the integration of carbon nanotube electrodes into an standard AMLCD process could be achieved without major changes to the individual steps. The complete production from the bare substrate up to driver bonding could be realized by the author himself with help from colleagues for color filter deposition. The deposition of the self-assembled monolayer by immersion in the aqueous solution had no detectable influence on TFT performance. A volume of 6 mL HiPCO p SWNTs was sprayed onto each back- and frontplane (substrate size 10 cm × 12 cm) resulting in a sheet resistance of about 1 kΩ/□ to 2 kΩ/□. CNN etching by RIE did not show any negative influence either. Both is not surprising as the backplanes pass through several wet and RIE processes during production.

Figure 6.18a shows an SEM image of a CNN patterned as sup-pixel electrode on a dummy backplane having only source-drain metallization and passivation with vias to the metal. These are the two interfaces the CNN gets in contact with on the backplane. The silane worked sufficiently well as adhesion promoter on the Si$_3$N$_4$ as well. After rinsing and resist strip no damage to the CNN was stated. Once again residuals of bundles are perceivable in the etched regions which did however not cause any shorts between pixels. On the frontplane the CNN was directly deposited on the planarization layer. As it is an organic material, no surface treatment is necessary to get sufficient adhesion. An optical microscope image of a fully processed backplane before

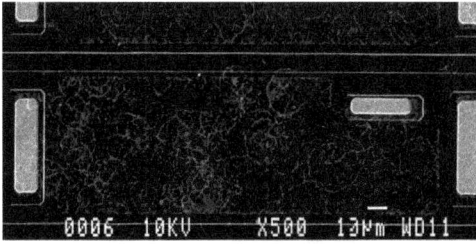

(a) SEM image showing the patterned CNN on a dummy backplane with only drain-source metallization and patterned Si_3N_4 passivation layer

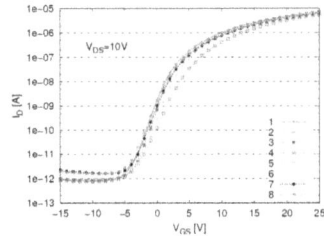

(b) Transfer characteristics of a-Si:H TFTs in the corners of AMLCD backplane; $S \approx 1.3\,V/dec$.

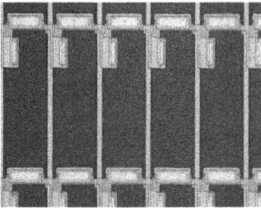

(c) Finished backplane with CNN electrodes

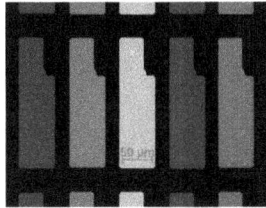

(d) Backlighted colorfilter substrate with black matrix

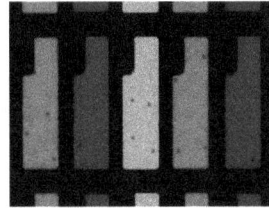

(e) Backlighted assembled AMLCD

Figure 6.18.: AMLCD process results

PI coating is shown in fig. 6.18c. Besides some minor dark spots, the CNN is quite homogeneous over the whole substrate size. The frontplane with RGB color filters and black matrix is shown in fig. 6.18d, a small section of a completed, backlighted AMLCD without polarizing films can be seen in fig. 6.18e. Besides spacer balls some texture from the CNNs can be perceived. Transfer characteristics of 8 a-Si:H TFTs in fig. 6.18b show that the achieved TFT performance fulfills the minimum demands of the first order approximation for I_{off} and I_{on} calculated in section 2.4.5. The subthreshold voltage of $S \approx 1.3\,V/dec$ is however rather poor, resulting in a higher gate voltage range.

A row driver and column drivers were bonded to each finalized display using a semiautomatic bonder. Bonding conditions and adapted tooling first had to be developed. The used column drivers in chip on flex (COF) package have a contact pitch of only $25\,\mu m$ at the interface to the LCD. For the $320 \times RGB = 960$ columns a total of 3 COFs had to be cascaded and bonded to display and PCB. Some missing columns are therefore not surprising. The used addressing system, based on a Field Programmable Gate Array (FPGA), was developed in a previous work [64]. The complete setup with power supply, FPGA board, PCBs for row and column drivers and module for DVI import is shown in fig. 6.19a.

Content can be generated by a pattern generator realized in the FPGA or by connecting a PC with PCI or VGA export. One of the produced displays with a test pattern is shown in fig. 6.19b. The display only shows some minor column and pixel defects. The column defects are caused by either broken buslines or insufficient driver bonding. The cause of the individual pixel defects is difficult to evaluate. They could be

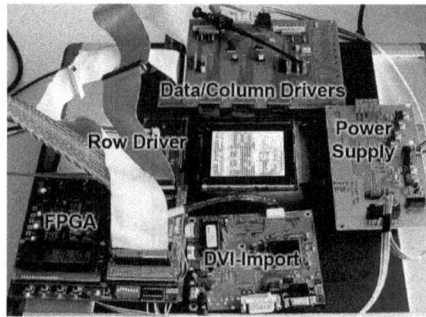

(a) Display with bonded drivers and addressing system.

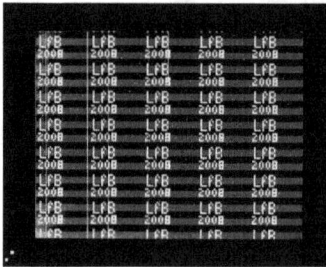

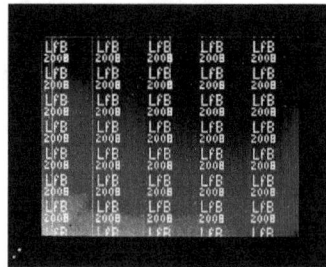

(b) On/off test pattern   (c) Test pattern with backround in grey scales

Figure 6.19.: Finalized AMLCD demonstrator

caused by defective TFTs or missing contact between pixel electrode and TFT. The number of defects is sufficiently low to exclude a general problem. The test pattern in fig. 6.19b does not contain any grey scales and is therefore less demanding. The sub-pixels only need to be rendered fully on or off. Figure 6.19c shows are more demanding test pattern where the blue background is supposed to show a continuous linear transition from bright blue in the bottom to black at the top. This pattern reveals some more unwanted behavior. The transition is not identical over the whole display width and there is a bright spot at the bottom left corner. Once again it is difficult to clearly determine the source of these defects. Any inhomogeneity over the display area could be caused by a gradient in the counter-electrode potential due to the elevated sheet resistance of the CNN. A possible workaround in this case would be to add a low resistance metal grid which is in contact with the CNN and hidden by the black matrix. Such a grid was tested on not fully addressed samples and did not seem to give any remarkable improvements. Other causes for the seen defects might be inhomogenous TFT performance caused by an uneven back channel etch or contamination from sources like process chemicals, PI layer, glue frame, etc. It also needs to be mentioned that finding the correct addressing signals for an AMLCD is quite difficult as parasitic capacitances in the matrix have a strong influence on the pixel voltages while a direct probing of the pixel voltage is not possible. A better definition of the external signals to be applied can only be gained by precise modeling and simulation of the complete matrix, which is not part of this work.

The overall quality of the realized displays is quite surprising given that the complete process chain from bare substrate to addressed display takes several weeks of mostly manual handling in a university lab environment and only a few samples were produced. Given the complexity of technology and electrical addressing, a huge effort would be necessary to optimize all steps to a degree where the influence of a single element could be judged in a more conclusive manner. The realized displays show that carbon nanotube networks can replace ITO pixel electrodes in a standard AMLCD process without major implications on the other production steps. Despite the reduced electro-optical performance, the CNT-AMLCDs work well with static and animated content, which is a successful proof of applicability of CNTs as transparent conductive film in AMLCDs.

6.5.4. Polymer organic light emitting diode

Organic light emitting diodes (OLEDs) are another display technology that became more and more important in recent years. While AMLCDs are light valves with a high intensity backlight behind, OLEDs are multi-layer thin-film devices that directly emit light. The advantage over their inorganic brothers is the ease to fabricate an extended, planar pn junction. This allows the emission area to be tailored as desired. Two basic concepts for creating full color displays are currently on the market. The base colors red, green and blue can either be created from color ermitters or by a white emitter plus color filters, similar to AMLCDs. The material stack usually consists of an ITO anode, an organic multi-layer stack that contains the emitting layer surrounded by electron and hole injection layers plus possible other organic layers that optimize transport and recombination of the charge carriers. The stack is finished by a low work function reflective metal cathode like Ca plus a reinforcement by e.g. Al. The organic emitter might be evaporated or deposited from solution. For this work, a polymer organic emitter was used in solution. Processing under inorganic atmosphere plus a high performance encapsulation of the final product are necessary to avoid damage caused by oxygen and water vapor.

An OLED is a current driven device, while LCDs are low current voltage driven devices. A low sheet resistance is consequently indispensable to avoid brightness gradients on larger size diodes. For larger area lighting applications, low resistivity ITO is even improved by adding metal grids. Considering the performance of the CNT-TCFs reported in section 6.2, they are not ideal to replace the ITO anode. A short study of feasibility was nevertheless realized with a colleague working on OLED technologies [49].

The CNT-TCF was spray coated on glass substrate, rinsed and patterned. Once finished, it was oven dried to minimize any humidity that might be trapped in the CNN. All further processing and testing was done under inert atmosphere. The organic layers where deposited by spin-coating followed by evaporation of the cathode through a shadow mask. Even though the results clearly need improvement, working OLEDs could be realized in the first iteration (see fig. 6.20). Anode and cathode are patterned in a way to realize OLED bars of different length as can be seen in fig. 6.20a. Adressing all bars in parallel results in a profound drop of liminance for the larger bars. The second bar from the top is much brighter than the others. Besides this, other defects are visible. There is a radial brightness difference that is caused by the spin-coating process.

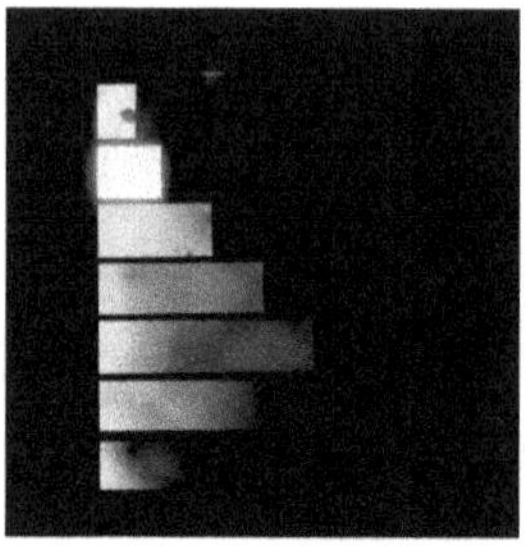

(a) Several OLED pixels of different size addressed in parallel.

(b) Best OLED pixel with size of 6 mm × 5 mm.

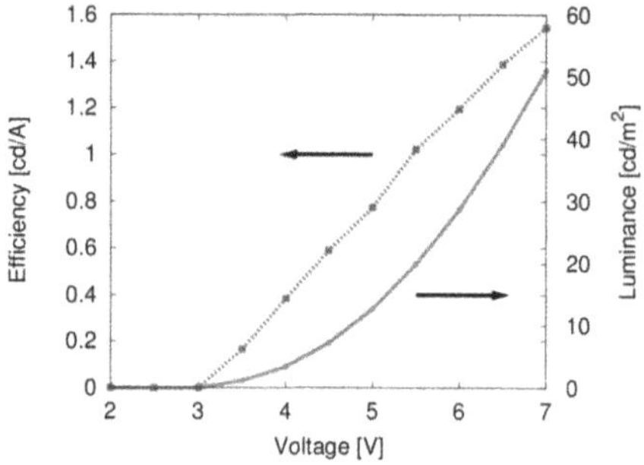

(c) Electro-optical characterization of the best OLED pixel.

Figure 6.20.: Organic light emitting diodes with carbon nanotubes networks as anode driven at 6 V.

All less well working bars have dark spots. These seem to have caused some short circuits resulting in a reduced brightness for these segments. As the OLED stack thickness is only in the range of some 100 nm, the topography of the CNN might be the cause of this. As expected, the long bars further show a gradient in brightness.

In fig. 6.20b only the best performing pixel with a size of 6 mm × 5 mm is supplied with power. This shows that at least for smaller size segments a homogeneous luminance can be achieved in absence of shorts. Efficiency and luminance of this segment are plotted in fig. 6.20c. At 7 V an efficiency of 1.55 cd/A and a luminance of 51 cd/m^2 were achieved. Following the trend of the curves reveals that these values can be higher with a more elevated driving voltage. These were however not applied to avoid possible damage to the OLED. Similar devices with an ITO anode and identical material stack reached - after a non-negligible process optimization - efficiencies of 4 cd/A to 9 cd/A and luminance values up to 1200 cd/m^2 [49]. No further resources were put in the optimization of these initial results. Summarizing, it can be stated that with some improvements CNNs could replace the ITO in applications where flexibility and repeated bending are more important than a low sheet resistance.

7. Carbon nanotube network thin-film transistors

This chapter describes the fabrication of thin-film transistors using carbon nanotube networks as semiconducting channel material. From initial tests on oxidized silicon wafers, the setup was quickly transferred to a process compatible with glass and plastic substrates. The envisaged application are flexible CNT-TFT based active-matrix displays. The experimental results stay however on a TFT level.

Several deposition methods for the CNN on TFT subtrates were tested. The spray coating process which yields good results in the realization of carbon nanotube TCFs, does not provide sufficient reproducibility and parameter control to repeatedly achieve homogeneous, low density networks which are necessary for a working semiconducting device [85]. A promising concept is the deposition of carbon nanotubes from solution by electrophoresis [128]. TFT substrates are immersed in a SWNT dispersion and individual nanotubes are attracted to the channel area by an AC electrical field between source and drain electrodes. The big advantage is a selective deposition in the channel area without further need for semiconductor patterning. While the deposition of nanotubes could be achieved in a controlled manner, the TFT performance was rather poor [129]. This is due to the fact that metallic nanotubes experience higher attraction forces than semiconducting ones. Even when using nanotube suspensions that contain mostly semiconducting SWNTs, the working devices still had poor performance.

The results presented below are all realized with CNNs deposited by the spin-coating process described in section 5.3 as it allows to deposit high quality nanotube networks in the necessary density with repeatable parameters. All processes were developed with the classical mix of 1/3 metallic SWNTs and 2/3 semiconducting SWNTs. Highly enriched semiconducting SWNT products were available only at the end of the research project.

7.1. Mask set layout used for CNT-TFT realization

Given the fact that a spin-coating process is used for deposition of the semiconducting layer, consisting of particles, a radial design shown in fig. 7.1a was developed [85]. It consists of 9 sectors, one for each TFT geometry. Each sector contains 10+7 radial TFTs (r-TFT) and 5+5 tangential TFTs (t-TFT) as indicated in fig. 7.1b. Indexing of individual TFTs starts from the circle edge. The nine different TFT geometries are

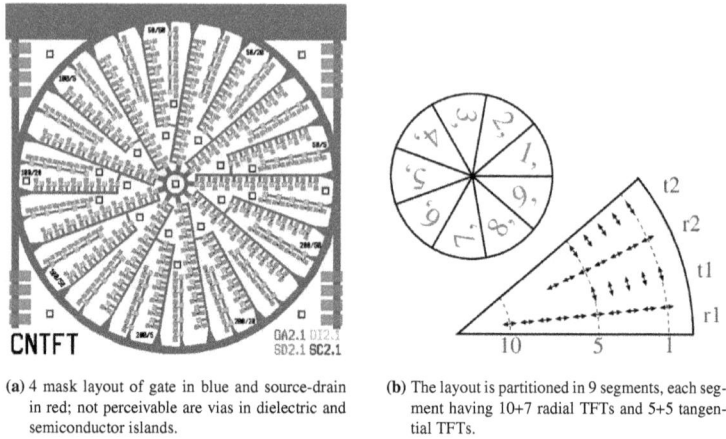

(a) 4 mask layout of gate in blue and source-drain in red; not perceivable are vias in dielectric and semiconductor islands.

(b) The layout is partitioned in 9 segments, each segment having 10+7 radial TFTs and 5+5 tangential TFTs.

Figure 7.1.: Radial TFT mask design for substrates of 50 mm × 50 mm.

the combination of channel length $L_C = 5\,\mu m$, $20\,\mu m$ and $50\,\mu m$ and channel width $W_C = 50\,\mu m$, $100\,\mu m$ and $200\,\mu m$. The complete gate structure is connected to the upper contact bar to allow the electro-chemical formation of the gate dielectric as will be described in the following section.

7.2. Process flow for CNT-TFT production

This section describes the TFT fabrication process that was developed for the realization of carbon nanotube thin-film transistors. It was designed to be a low temperature process that can be used on both glass as well as on plastic substrates. In the case of a glass substrate Corning® display grade Eagle 2000™ is used. A polyethersulfone (PES) based foil from Sumitomo Bakelite Co., Ltd. is used as plastic substrate. The full process details are listed in appendix A.6.

As the topography of the TFTs is bottom gate and top contact, the first step is the realization of the gate electrodes and connecting buslines. This is done by the deposition of a 150 nm thick Al layer by sputtering and subsequent patterning by photolithography and wet etching (see fig. 7.2a). In the case of PES foils a thin Ta_2O_5 layer is deposited before the Aluminum (see section 5.1.1). This is done in the same tool without breaking the vacuum and thus does not significantly increase the process complexity. The purpose of the thin oxide layer is to increase the adhesion of metal layers on the plastic. Additionally it serves as an etch-stop layer in dry-etching processes. The oxide layer might in the case of AMLCDs also serve as barrier against water vapor. The thickness was kept at 30 nm, which results in a continuous layer that is still thin enough to be flexible. Also, there is no additional surface tensions that would bend the substrate.

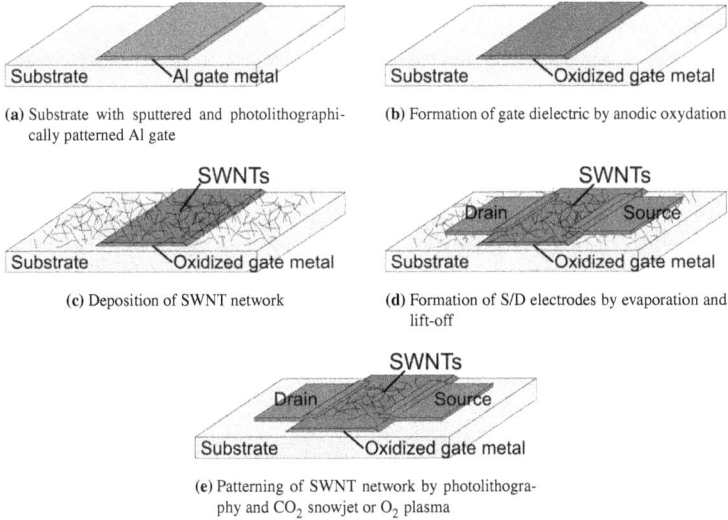

(a) Substrate with sputtered and photolithographically patterned Al gate

(b) Formation of gate dielectric by anodic oxydation

(c) Deposition of SWNT network

(d) Formation of S/D electrodes by evaporation and lift-off

(e) Patterning of SWNT network by photolithography and CO_2 snowjet or O_2 plasma

Figure 7.2.: Fabrication process of CNT-TFTs

The dielectric layer is formed by anodic oxidation of the Al patterns in hydrogen peroxide (H_2O_2). This process was developed and optimized in a preceding research project [36]. The substrate is put into the H_2O_2 solution while applying a voltage between the gate metal, which serves as anode, and a blank metal cathode which is placed in a distance of a few millimeters. The voltage applied between anode and cathode leads to a high electrical field in the thin native aluminum oxide (Al_2O_3). This electrical field forces hydroxy ions (OH^-) to migrate through the oxide layer towards the metal and Al^+ ions to migrate in the other direction. At the two interfaces (Al/Oxide, Oxide/solution) both types of ions react to form the Al_2O_3. This way the oxide layer grows in both directions. Vias through the dielectric are formed by applying photoresist patterns before the oxidation process. A well formed oxide is achieved by keeping a constant current density (in this case $0.35 \frac{mA}{cm^2}$) during the growth process. The final thickness of 60 nm is determined by the maximum voltage V_{max} that is allowed between anode and cathode by voltage regulation and can be calculated by the following equation:

$$t_{oxide} = V_{max} \cdot 1.73 \frac{nm}{V}. \tag{7.1}$$

Since Al is attacked by both acidic and basic solutions, the pH value has to be kept close to 7. This is done by buffering the solution with ammonium hydroxide (NH_4OH). The current driven process is self-healing. A pin-hole in the oxide layer leads locally to a higher current density, resulting in a homogenization of all irregularities. Also the side walls of the metal patterns are well oxidized. Once V_{max} is reached the current decreases exponentially. A short period of formation (90 s), once the current is going down, assures a homogeneous layer formation. The photoresist patterns for via creation are stripped afterwards. Figure 7.2b shows the resulting oxide layer in red. In reality the grayish color of the Aluminum changes only slightly.

Before spin-coating deposition of nanotubes, the self-assembled monolayer (SAM) of 3-Amino-propyl-triethoxy-silane (APTS) is first deposited by soaking the substrates in a 1% aqueous solution for 1h plus rinsing and blow-dry afterwards. The APTS monolayer provides amine-terminated end groups that give good adhesion to SWNTs [94] (see section 5.1.2). The formation of such SAMs is known to work well on SiO and showed also to be effective on Al_2O_3, Ta_2O_5 as well as on glass. The nanotubes are then deposited by the spin-coating approach which is described in section 5.3 and further discussed in section 7.3. The result is a full area deposition as shown in fig. 7.2c.

The source and drain contacts are formed on top of the nanotube network by lift-off. A negative photoresist is first deposited and dried. The S/D patterns are transferred by a masked UV-exposure. The UV radiation crosslinks the exposed negative resist and makes it insoluble for the developer. The combination of crosslinking from the top and over-development leads to hanging edges of the resist as indicated in fig. 7.3. The metal layer is deposited onto this resist pattern by thermal or e-beam evaporation. In the high vacuum

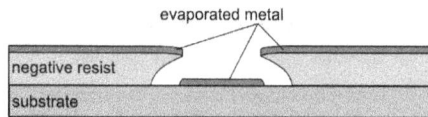

Figure 7.3.: Evaporated metal on negative resist with hanging edges for optimized lift-off.

the evaporated metal vapor is propagating directly from the source to the substrate without collisions with gas molecules. This directed material transport leaves the borders below the hanging resist edges uncovered. The resist is removed by soaking the substrates in a solvent and the metal above is lifted, leaving metal patterns with well defined edges (see fig. 7.2d). The nanotube network is well covered by the metal, leading to a good electrical contact. Another advantage of the evaporation process it the low kinetic energy during the deposition. Sputter deposition would damage the nanotubes.

The TFT fabrication process is finalized by patterning of the nanotube network. As with other semiconductors, this patterning can significantly decrease lateral leakage currents that would otherwise influence the off-current of the TFT and therefore diminish the on/off-ratio. A resist pattern is created by standard lithography. A short O_2 plasma is sufficient to damage or completely remove the nanotubes from the surface. During this etch step the channel areas are covered by the resist pattern and also the source/drain metalization serves as mask. After the plasma etch the resist can be stripped or kept as protection. It does however not serve as efficient passivation layer.

Images of finalized TFT substrates are shown in fig. 7.4a (glass substrate) and fig. 7.4b (flexible PES foil). An optical microscope image of a single TFT before CNT deposition is shown in fig. 7.4c. The described process can be applied to both glass and plastic substrates without changes to the process flow. TFTs realized on the flexible PES foil can be bent to a radius of at least 1 cm without implications to material integrity or TFT performance. The adhesion of the Pd source-drain metallization on the Ta_2O_5 layer on PES also proved to be sufficient by scotch tape test. The TFT substrates are much more flexible than the display demonstrators presented in section 6.5.2 as they consist of a single substrate and no polarizing filters are laminated.

(a) Glass   (b) Flexible PES foil

(c) Single TFT with $L_C = 50\,\mu m$ and $W_C = 100\,\mu m$;
source, drain and gate contact pads are indicated.

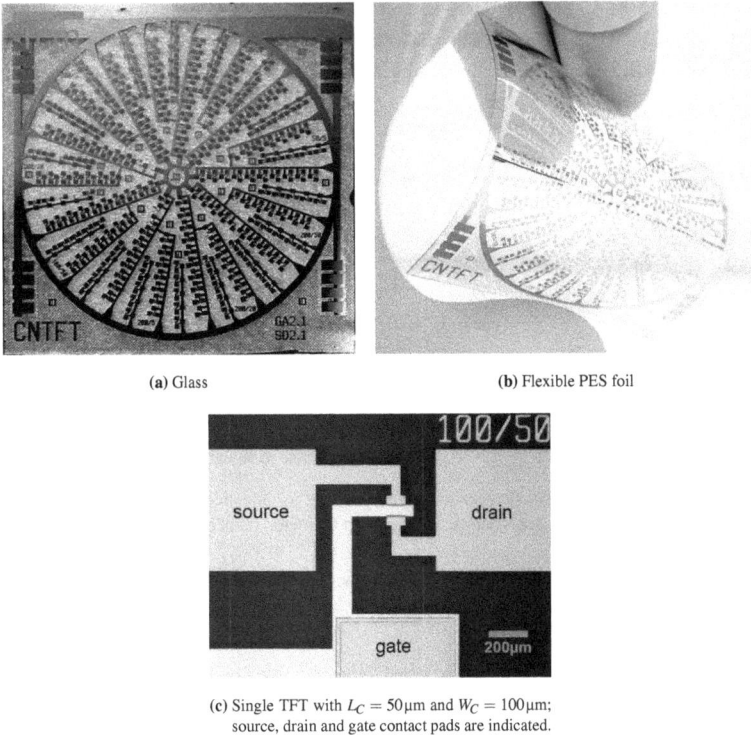

Figure 7.4.: Finalized TFT substrates

7.3. Optimization of the spin-coating process

Initial tests with two manually manipulated syringes resulted quickly in first functional TFTs. As the density of the CNN turned out to be a very important factor for the TFT performance, a precise control of the spin-coating process is crucial. The setup shown in section 5.3 was developed and optimized in several iterations [59, 72, 85]. Another important factor is the interaction of the used chemicals.

7.3.1. Dispensing of SWNT suspension and solvent

In the spin-coating process the mixing liquids are supposed to cover the complete substrate surface in order to achieve a deposition of SWNTs in all the TFT channels. The amid-terminated SAM creates however a hydrophobic surface. Given that the deposition is done from an aqueous suspension, the coverage of the substrate was investigated in more detail [59]. The deposition is most important in the TFT's channel area, which is covered by Al_2O_3. Dummy glass plates, fully covered with Al were oxidized and coated with the SAM. Figure 7.5a shows an insufficient wetting during spin-coating. Only the center is fully covered, towards the edges the liquid flow retracts to narrow channels. The resulting deposition shown in fig. 7.5b

(a) Snapshot during coating

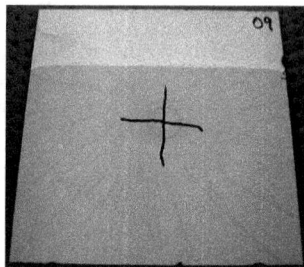

(b) Traces after deposition and drying; image contrast strongly increased; black center cross serves only as camera focus point.

Figure 7.5.: Insufficient wetting of suspension/solvent mixture on gate dielectric covered with APTS SAM. Both suspension and solvent use the same tip diameter, dispense pressure and flow rate.

clearly confirms the previous finding. The different channels obviously don't change once they are created. The spinning speed only has a very minor influence on this behavior.

Several solvents with different viscosity were tested. They are listed by increasing viscosity in table 7.1. Methanol is the initially used solvent. While Isopropanol with the highest tested viscosity results in the

Table 7.1.: Solvents tested for SWNT deposition by spin-coating.

Solvent	Viscosity [mPa · s, 25 °C]
Acetone	0.295
Methanol	0.545
Ethanol	1.074
Iso-Propanol	2.100

best coverage, it is not efficient enough in removing the micelles from the CNTs and almost no SWNTs are deposited. All three other solvents were not able to cover the complete substrate with the initially used dispense system.

Sufficient coverage of at least the used 5 cm x 5 cm substrates was achieved by increasing the volume flow of the solvent. In order to keep a well controlled vertical mixed stream, a bigger diameter dispense tip was chosen for the solvent. The pressure of the suspension stream is adapted to achieve the necessary deflection of the resulting intermixed stream. Good results were achieved with the parameters noted in table 7.2.

Table 7.2.: Spin-coating parameters for CNT deposition.

Spinning speed	3000 rpm
Solvent	Methanol
Solvent tip diameter	0.01 inch
Solvent pressure	10 psi
Suspension tip diameter	0.006 inch
Suspension pressure	Adapt to have vertical intermixed stream

7.3.2. Static vs. moving dispense head

With the optimizations explained above, the coverage of the CNN is largely improved and the deposition can be controlled more precisely. The TFT performance showed nevertheless big spreading even once 98 % s-SWNTs could be purchased and deposited. It turned out that there is always some radial dependency of the network density D. This becomes already visible with the center circle in fig. 7.5b. A static dispense unit placed at the center of the spinning axis is therefore not a good solution for homogeneous results. The dispense head was consequently mounted on the linear bearing shown in fig. 7.6. This allows a linear motion during dispensing. The motion axis is aligned to go through the spinning axis. The process sequence is:

1. start spinning of substrate
2. start solvent dispensing in center of substrate
3. add suspension dispensing with programmed duration
4. manually move dispense head back and forth in constant speed; point of return at the substrate edges
5. once suspension timer has finished, stop solvent dispensing after a few seconds and slow down spinning speed.

With this modification the homogeneity of CNT-TFT performance is highly improved. The results are presented in section 7.5.

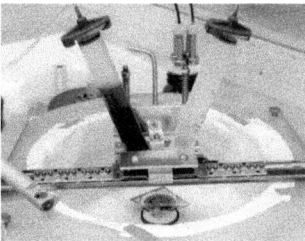

(a) Dispense head attached to linear bearing.   (b) Streams mixing above spinnig substrate.

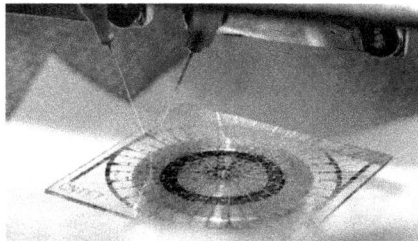

Figure 7.6.: Spin-coating with linear deplacement of dispensing unit.

7.4. CNT-TFT performance with a 1:2 mix of m-swNTs and s-swNTs

The simulation results presented in chapter 3 show that even though the classical mixture of SWNT feed-stock contains 1/3 of metallic nanotubes, an overall semiconducting CNN can be created. Functional devices could be achieved quite early on oxidized silicon wafers with the doped Si wafer as back-gate. The main effort then shifted towards transferring the process chain to materials compatible with display technology, with a special focus on flexible applications. With the mask set from section 7.1 and the developed process flow, presented in section 7.2, functional devices were realized on display-grade glass and plastic substrates. For each experimental series the deposition duration t_{dep} of the spin-coating process is varied between several substrates to find the optimum network density. In this thesis the characterization of the network density is only done indirectly by TFT characterization at the end of the process chain. A quantitative characterization of the network itself would only be possible by means of AFM measurements. Even with such a high precision microscope, the detection of individual nanotubes proved difficult. With scan sizes in the range of $10\,\mu m$ square, a mapping of several positions on a single substrate would already take half a day or more.

Figure 7.7a shows a scatter plot with two point clouds of extracted on/off-ratio vs. device charge carrier mobility μ_{device} for a range of TFTs (containing all types of geometries listed in section 7.1), realized on glass and plastic. A non-negligible number of non-functional devices (open-circuit, short-circuit, detected slope not corresponding p-type TFT) were eliminated during parameter extraction. Considering the fact that fig. 7.7a shows a double-logarithmic plot, it becomes obvious that the performance spread is enormous. Many devices achieve on/off-ratios of 5 orders of magnitude. On the other hand there are device charge carrier mobilities that go up to $10\,cm^2/(V\,s)$. Finally there are only very few devices that achieve $I_{on}/I_{off} > 10^4$ combined with $\mu_{device} \approx 1\,cm^2/(V\,s)$, which is the charge carrier mobility that can be achieved with a-Si:H and some organic semiconductors. A significant difference between TFTs realized on glass and plastic substrates cannot be perceived. The fact, that the highest on/off-ratios were achieved on glass could be due to the fact, that an optimal network density was not achieved on the fewer plastic substrates.

Two example transfer characteristics of a TFT with high I_{on}/I_{off} on glass and on plastic are given in fig. 7.7b. Off-currents and on/off-ratios are in a range that start to be interesting for AMLCD applications. The device mobilites are however almost two orders of magnitude smaller than $1\,cm^2/(V\,s)$, which can be achieved by standard a-Si technology.

Even though fig. 7.7a shows results of different channel geometries and varied deposition times, no optimum parameter set can be determined that would narrow down the performance spread on a single substrate for a defined geometry. When classifying the plot by channel length L_C (see fig. 7.7c), some trends however become visible. The behavior well reflects the network density regimes defined in section 2.1.4. The area marked with (2) corresponds to D just above p_{mix}. There are semiconducting paths between source and drain but the network is still sparse. As μ_{device} is calculated in reference to the channel area it has a small value. With higher D (following the colored lines towards (3)) the channel can transport more current, increasing I_{on} and consequently the on/off-ratio as I_{off} still stays low. At a certain point D reaches p_m, indicated by (4). The mobility gets even higher but the on/off-ratio collapses due to the metallic connections between source

(a) On/off-ratio vs. charge carrier mobility of a range of CNT-TFTs realized on glass and plastic.

(b) Transfer characteristics of TFTs with high on/off-ratio on glass and plastic; S: sub-threshold swing.

(c) Data from (a) classed by L_C.

(d) Data from (a) classed by W_C.

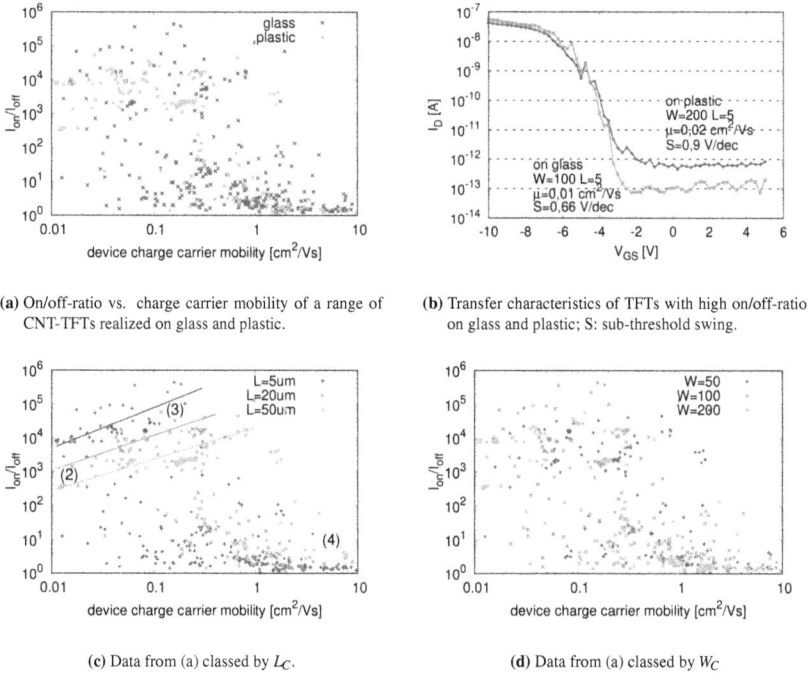

Figure 7.7.: CNT-TFT performance with mix-SWNT and static, centered dispense unit, $V_{DS} = -0.1$ V.

and drain. See also I_{off} and I_{on} vs. μ_{device} in fig. D.1, appendix D. While I_{on} rises linearly with μ_{device} in the double log plot I_{off} stays in the pico-Ampere range before it rises quickly with a threshold depending on channel geometry.

Besides these general trends, the classification by L_C, further marked by the colored trend lines, leads to two important findings:

1. Devices with shorter L_C can achieve higher on/off-ratios for a given μ_{device}
2. For devices with shorter L_C the on/off-ratio collapses at lower μ_{device}.

The first finding can be explained with the tube-to-tube barriers in the network. With increasing L_C each current path has more inter-tube connects, which according to section 2.1.4 cause a significant resistance increase compared to the intra-tube resistance. The second finding can already be concluded from simulation results in chapter 3. A higher aspect ratio channel (W_C/L_C) often has a lower CNT active-ratio, which means that the electrically active part of the CNN is not covering the complete channel area resulting in a smaller μ_{device}. Once the channel starts being bridged by metallic connections the mobility increases at low on/off mainly due to a high I_{off}. The same scatter plot with classification by W_C is shown in fig. 7.7d. The point clouds are less well regrouped in this case. The plot shows however that high on/off-ratio combined with high μ_{device} are more likely for narrower devices.

7.5. CNT-TFT performance with 98% semiconducting SWNTs

Once highly semiconducting mixtures became commercially available, they were deposited on the same types of materials and with the processes described above. The used IsoNanotubes-S (see table 4.1) are specified to contain 98 % of purely semiconducting SWNTs, dispersed in a proprietary ionic surfactant mixture at a concentration of 0.01 mg/mL. In a first iteration the s-SWNTs were spin-coated with similar deposition parameters as used before for self-made suspensions [72]. Even though some TFTs with $I_{on}/I_{off} > 10^5$ and $\mu_{device} > 1\,cm^2/(V\,s)$ could be achieved right away, a similar performance spread as for mix-CNNs can be seen in fig. D.2a in appendix D. The individual data points for a fixed deposition duration are once more widely spread. When looking only at the substrate which contains the highest performing TFTs and classing the data points by TFT position, the source of the large spread becomes more clear (see fig. D.2b). Devices that are close to the substrate center all have high mobility, but low on/off-ratio. Better performing devices can only be found towards the outer half of the substrate. A clear winner is however not visible either.

The dispense unit was subsequently changed and the head was mounted on the linear bearing presented in section 7.3.2. The resulting plot of on/off-ratio vs. μ_{device} for 6 substrates with different s-SWNT deposition times are given in fig. 7.8a. These point clouds are much better regrouped. With a deposition time of 10 s D is not sufficient and leads to a high spread in μ_{device} and on/off-ratio. The three lines corresponding to the three different L_C (compare fig. 7.7c) are well perceivable. For the following deposition durations t_{dep} the spread in mobility is much smaller. A very crucial improvement in homogeneity of device performance can be observed when only looking at devices with $L_C = W_C = 50\,\mu m$ of the sample with $t_{dep} = 20\,s$. The corresponding data points are marked with black circles in fig. 7.8a. For theses devices an average on/off-ratio of 5 orders of magnitude and μ_{device} in the range of $4\,cm^2/(V\,s)$ to $6\,cm^2/(V\,s)$ can be achieved. Transfer and output characteristics of such devices are given in figs. 7.8b and 7.8c, respectively. Although there is still some spreading in the threshold voltage, all 10 transfer curves for $V_{DS} = -0.1\,V$ are quite closely regrouped with a constant off-current in the pA-range. The gate voltage range of $\pm 10\,V$ is relatively low. The similarity of the 10 transfer curves also shows that the network density is quite homogeneous with the improved deposition technique as all 10 TFTs are located on the same substrate with different radial positions. The output characteristics in fig. 7.8c reveal a very good contact between CNN and source and drain contacts with no perceivable diode effect and some reasonable saturation. The Schottky barrier between Pd contacts and CNN therefore seems to be quite low. TFTs with $L_C = 50\,\mu m$, $W_C = 100\,\mu m$ and $L_C = 20\,\mu m$, $W_C = 50\,\mu m$ are also grouped around the same point with slightly less on/off-ratio and two more outliers (not indicated in fig. 7.8a).

Figure 7.8b however also shows that for higher drain-source potentials the off-current increases. While at $V_{DS} = -1\,V$ the on/off-ratios still increase towards 10^6, at $V_{DS} = -10\,V$ the off-currents are already in the

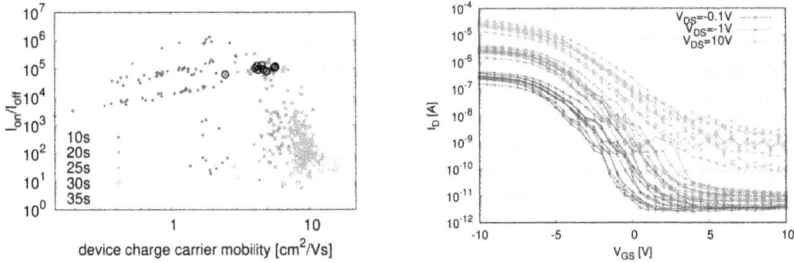

(a) On/off-ratio vs. μ_{device} for $V_{DS} = -0.1$ V, black circles mark TFTs with $L_C = W_C = 50\,\mu$m and 20 s deposition.

(b) Transfer characteristics for 10 TFTs with 20 s deposition duration and $L_C = W_C = 50\,\mu$m.

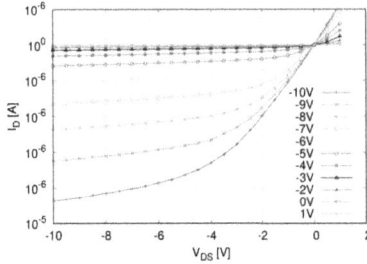

(c) Output characteristics of one TFT from (b), V_{GS} as parameter.

Figure 7.8.: CNT-TFT performance with 98 % semiconducting SWNTs and linear dispense unit.

nA-range. This is much too high for an AMLCD application (see values in section 2.4.5). On the substrate with $t_{dep} = 10$ s the off-current stays in the same pA-range for all 3 values of V_{DS} combined with on/off-ratios $> 10^5$ (see fig. D.3). The transfer characteristics are however less well regrouped with more spread in on-currents and therefore much more variation in μ_{device} (compare fig. 7.8a). Table 7.3 gives values for the subthreshold slope of TFTs on both substrates at different V_{DS}. For comparison, the a-Si:H TFTs presented in section 6.5.3 have $S \approx 1.3$ V/dec at $V_{DS} = 10$ V. At lower t_{dep} the subthreshold swing is much smaller. The lower network density devices therefore seem to have significantly less charge carrier traps.

Table 7.3.: Subthreshold swing for samples with t_{dep} of 10 s and 20 s and different V_{DS}; average value for 3 transfer characteristics each.

V_{DS} [V]		-0.1	-1	-10
S [V/dec]	10s	0.38	0.42	0.58
	20s	1.24	1.12	1.78

When increasing deposition times above 20 s the on/off-ratios start to degrade rapidly, demonstrating that even with highly enriched s-SWNTs the network density needs to be carefully controlled. Assuming that a purity of 98 % s-SWNTs would leave 2 % m-SWNTs in the suspension, a much larger process window should be available when only looking at the simplified stick percolation model. According to section 3.2.4 there should be a factor of $1/0.02 = 50$ between $p_{50\text{mix}}$ and $p_{50\text{met}}$. A linear correlation between D and deposition duration can be presumed. The values of table 3.1 in section 3.2.4 can therefore in first approximation be interpreted as deposition times. In the case of the β-distribution this would mean that mix-CNNs are always percolating above 4.4 s. Shorting by m-SWNTs would then only appear at $t_{\text{dep}} > 150$ s. It is not possible to read from fig. 7.8a at what deposition time percolation of the mix-CNN begins. It can however be concluded that D for $t_{\text{dep}} = 10$ s is above $p_{99.5\text{mix}}$ given that 89 out of 90 TFTs are working. The spread in performance shows further, that the CNT active-ratio is still below 1. The example given above therefore seems close to reality and it is surprising, that the off-currents rise already drastically for a deposition duration of 25 s.

Figure D.4 shows on- and off-current vs. μ_{device} for the discussed TFTs. While the on-currents increase linearly with D in the double-log plot, the off-currents stay at a low level and mount rapidly (for $L_C = 5\,\mu\text{m}$ sooner as for $L_C = 20\,\mu\text{m}$ and $50\,\mu\text{m}$). A significant doping by O_2 or other molecules that would shift the fermi level of the s-SWNTs into the valence band and rend them metallic seems unlikely, as it would also affect devices with lower network density. A denser CNN might however be more prone to ionic contamination from e.g. residual surfactant. Another explication could be a shielding effect where s-SWNTs farther away from the interface to the dielectric might not be completely depleted by the gate potential any more. A denser network could also lead to more pronounced bundling which might result in a smaller or collapsed effective bandgap of resulting bundles. The exact reason for quickly rising off-currents with increasing D of 98 % semiconducting SWNTs is however difficult to derive from these data.

Another serious issue is a strong hysteresis that becomes visible when measuring transfer characteristics where V_{GS} is swept from negative to positive voltages and back. The result for TFTs on the two substrates with t_{dep} of 10 s and 20 s are shown in fig. 7.9a. Such a behavior is often attributed to trapped charges, either in the dielectric or at the interface to the semiconductor. Despite a significantly lower S for the 10 s devices the hysteresis is similar for both network densities. A significant influence seems already to be attributed

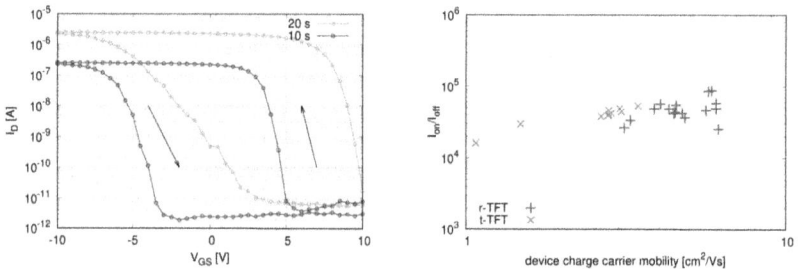

(a) Hysteresis depending on scan direction of V_{GS};$L_C = W_C = 50\,\mu\text{m}$.

(b) Performance depending on channel orientation in tangential or radial direction; $L_C = 50\,\mu\text{m}, W_C = 100\,\mu\text{m}$.

Figure 7.9.: Observed non-ideal behavior of CNT-TFTs.

from the dielectric itself. The same technology was already used for the realization of organic TFTs that also exhibited a similar hysteresis. A more profound analysis of the dielectric revealed that the low temperature process indeed yields to some charge carrier trapping in the oxide layer [63]. CNT-TFTs realized on silicon wafers however also showed a similar hysteresis behavior. The influece of the source drain metallization was investigated by realizing TFTs with sputtered Au bottom contacts on the same base substrates. Similar TFT performance compared to the usual Pd top contacts was achieved with slightly lower mobility values (see fig. D.5a in appendix D). The hysteresis stays however comparable (see fig. D.5b). It seems therefore likely that the source of the hysteresis is a combination of imperfect dielectric combined with trapping states at the dielectric/CNN interface or in the nanotube network itself. This is not surprising given the topology of the CNN. All atoms of a SWNT are exposed to it's surroundings. Even though SWNTs are inert and have in absence of defects no dangling bonds, molecules attached by Van-der-Waals forces can have an influence on the carrier density (doping effects) or the band structure itself. In randomly oriented CNNs the crossings between indiviudal tubes further have a strong effect on the electrical behavior of the final device. Given that these tunnel barriers and Schottky contacts are also quite exposed to the environment, charge trapping and hysteresis can be caused at these points. In all transistor technologies, the purity of the semiconductor and an effective passivation of the semiconductor from external influences are prerequisites for reliable device performance. Water adsorbed at the surface of dielectric and nanotubes was for example reported to create significant hysteresis in ambient air [70]. In the cited publication storage and characterization in vacuum proved to diminish the hysteresis.

An unwanted side-effect of the spin-coating deposition is a channel orientation dependent TFT behavior. This holds especially for the mobility values (see fig. 7.9b). It is known from literature that spin-coated SWNTs have a preferential orientation in radial direction [107]. This leads to a higher μ_{device} for channels in radial orientation while the on/off-ratio changes only slightly. This orientation-dependent characteristic seems to be difficult to compensate with the employed deposition process and might be a limiting factor when trying to realize circuits or matrices that cannot be organized in a radial orientation.

8. Conclusions and outlook

In the frame of this thesis the use of randomly oriented networks of SWNTs as transparent pixel electrodes and as semiconducting film in thin-film transistors was successfully demonstrated. The necessary processes from creation of suspensions to fabrication of complete display demonstrators and individual TFTs were mostly developed in the frame of this research project. Specific contact resistance characterization between different metals and CNNs are a fundamental building block to allow the integration of CNNs in display systems.

Besides the experimental part, a Monte-Carlo simulation software was developed to get a better under- standing of the percolation behavior and the topography of CNNs (see chapter 3). For the application of transparent conductive films these simulations are of only limited use. Dense networks, far above the perco- lation threshold, are deposited for the realization of such films. The main requirements are a highest possible conductivity at on optical transmission between 80 % and 90 %, which is best investigated by experiment. For the application of CNT-TFTs these simulations are however an important tool. The characterization of the network topography in fabricated TFTs is difficult to realize, even with methods like atomic force microscopy. A direct correlation between network density in simulation and experiment, combined with resulting TFT performance, was therefore not possible. Findings from simulation correlate nevertheless well with experimental data. The simulation software interprets the CNN as a stick percolation system. For the low network densities used in CNT-TFTs this seems a viable approach. The following general insights could be gained by simulation for the case of mixed CNNs:

- The ratio of TFT channel length to nanotube length should be $L_C/L_t > 5$ for SWNTs with uniform length and $L_C/L_t > 10$ for a length distribution similar to the one specified for IsoNanotubes-S.
- The CNT active-ratio (# of deposited CNTs / # of electrically active CNTs) should be close to 1 to have homogeneous TFT performance and a high μ_{device}.
- A channel aspect ratio $L_C/W_C \geq 1$ gives a bigger process window.

Transparent conductive films were fabricated from several types of SWNTs and a DWNT sample. The best performance was achieved with purified HiPCO SWNTs. They achieve an optical transmission of 86.4 % at 737 $\Omega/\square$ which is comparable to other literature values for wet deposited CNN-TCFs. The created networks tend to form bigger bundles, creating an interwoven network. This leads to an elevated surface roughness and might further absorb more light than a homogenous layer. A finer atomised spray might therefore lead to a smoother layer and possibly higher electro-optical performance. The best performing CNN-TCFs found in literature have a sheet resistance of 60 $\Omega/\square$ at 91 % optical transmission [43]. These are however doped by strong acids that are often toxic and volatile and were therefore not used in this work. Compared to sputtered ITO these best performing CNN-TCFs still have a sheet resistance that is almost one order of magnitude higher at identical transmission. In terms of electro-optical performance CNT-TCFs are therefore not able to rival the well established ITO. They might however be the better choice for highly flexible displays as they can withstand bending, twisting and stretching way better than ITO that fails quickly under such conditions [133]. Another advantage of CNNs over ITO is their lower refractive index. Narrow buslines down to 5 μm can be formed that stay continuous over long distances. Despite the roughness of the film a liquid crystal alignment layer could be formed on-top of the network that allows the realization of TN-LCDs. Well working segmented PDLCDs and TN-LCDs were realized on glass and on foil. The integration of the nanotube TCFs into a standard a-Si:H AMLCD process without any major modifications to the production flow was successfully demonstrated.

For the realization of CNN-TFTs it can be concluded that a fabrication process on glass and flexible plastic substrates was established. Several improvements of the deposition by spin-coating helped to achieve better repeatability and homogeneity in TFT performance. Even with classic CNT mixtures that contain 1/3 metallic SWNTs, working devices with good electrical performance can be realized. Finally, it is the combination of the improved deposition with nowadays available highly semiconducting CNT feedstock, which makes it possible to achieve device charge carrier mobilities μ_{device} of $4\,cm^2/(V\,s)$ to $6\,cm^2/(V\,s)$, combined with on/off-ratios of 5 to 6 orders of magnitude. The calculated device charge carrier mobility is dependent on the channel coverage. It is however a good measure for comparing performance and footprint of different TFT technologies. The findings from Monte-Carlo simulations could be verified by experiment. For short and wide channels, the TFT yield is comparably low. The devices still work but the on/off-ratio collapses quickly due to increased off-currents. Shorter channel lengths however resulted in higher on/off-ratios that can go up to 10^6. This proofs that conduction through nanotube networks is limited by the inter-tube barriers.

Most homogeneous results are achieved for $L_C = 50\,\mu m$ and a W_C of $50\,\mu m$ and $100\,\mu m$. For the implementation into an active-matrix, such a footprint is quite big, leading to a low resolution or low pixel aperture ratio. The more homogeneous TFT performance after process optimization reveals several issues of the fabricated CNT-TFTs. Charge traps in dielectric, CNN and their interface cause a significant hysteresis and a high subthreshold swing. The use of low temperature processes, compatible with plastic substrates makes these problems more difficult to master as high temperature annealing to evaporate surface-bonded H_2O or other contaminants is not possible. High quality and high-K dielectrics would further reduce parasitic effects. A trade-off between higher density networks with maximized μ_{device} and lower density ones with reduced subthreshold swing and better off-currents needs to be found. Even with 98 % semiconducting SWNTs, the process window between reduced CNT active-ratio and increased off-currents is however quite narrow. More profound research would be necessary to better understand the source of unintentional network conductivity for higher network densities and hysteresis effects. Conditioning and passivation methods that are common in all transistor technologies would need to be investigated to achieve more stable TFT performance. The ultimate goal would be the realization of dense, aligned networks between source and drain. In combination with n-type doping, complementary driver circuits could be realized as well. The achieved performance of CNT-TFTs could be sufficient to use them in AMLCD backplanes. The control over off-currents and device to device homogeneity however must be improved.

A. Process parameter

A.1. Cleaning of substrates

A substrate label is scribed on the backside of each substrate with a diamond tip before cleaning. Plastic substrates have a protection film that needs to be removed before processing. No additional cleaning is done. Glass substrates are cleaned with the following process:

- DI high-flow (10 min)
- DI-Hot with Mucasol; ultrasound (10 min)
- DI high-flow (10 min)
- DI-Hot; ultrasound (10 min)
- 80 °C DI H_2O which is slowly drained.

A.2. Transparent conductive film process

This process flow is compatible with glass and plastic substrates.

1. Label scribing on backside
2. Washing in case of glass substrates
3. Sputter deposition of etch-stop layer and metal contacts

3.1 Carrier	Material: Al	Process: DC	
pre-conditioning	Sputter etch	Power: 10 %	t = 2min
		Oszillations: 3	v = 0.57m/min
3.2 Etch-stop layer	Material: Ta_2O_5	Process: HF	thickness = 30nm
	Sputter etch	Power: 10 %	t = 1min
	Program: 6	Oszillations: 7	v = 0.4m/min
3.3 Metal contacts	Material: Au	Process: DC	thickness = 50nm
	Sputter etch	Power: 10 %	t = 2min
	Program: 3	Oszillations: 2	v = 1m/min

4. Photolithography metal contacts

4.1 Resist coating	Resist: AZ 650-TFP		
	Max speed	2500 rpm	
	Hotplate:	T = 120°C	t = 90s
4.2 Exposure	t = 3.8s	see product sheet for dosage	
	Mask: Metal contacts		
4.3 Development	Developer: TMA238WA 7:1 H_2O (ca. 200ml)		
	t = ca. 20s	1min rinsing in DI high-flow	
4.4 Wet etching	Au etchant @ room temperature		
	Sputter etch	Power: 10 %	t = 1min
	Duration ca. 30s	1min rinsing in DI high-flow	

| 4.5 Resist strip | Ethanol 1+2+3 in Ultrasound bath at roomtemp |
| | 10min, 10min, 5min, N_2 blow-dry directly after bath 3 |

5. Deposition SWNTs

5.1 Sputter etch	20%, 2min
5.2 SAM deposition	Material: 3-Amino-propyl-triethoxy-silane
	2ml in 200ml H_2O, immersion during 1h
	1min rinsing in DI high-flow
5.3 Spray-coating	HiPCO p suspension
	2ml on each 50mm x 50mm substrate
	6ml on each 100mm x 120mm substrate
5.4 Rinsing of	1. Beaker with H_2O at RT t = 10min
surfactant	2. Di high-flow t = 10min
	3. N_2 blow-dry + oven dry 15min @ 120°C

6. Patterning of SWNT layer

6.1 Resist coating	Resist: AZ 650-TFP
	Max speed 2500 rpm
	Hotplate: T = 120°C t = 90s
6.2 Exposure	t = 3.8s see product sheet for dosage
	Mask: CNT-TCF frontplane or backplane
6.3 Development	Developer: TMA238WA 7:1 H2O (ca. 200ml)
	t = ca. 20s 1min rinsing in DI high-flow
6.4 RIE	p: 0.06mbar Ar 10ml/min, O_2 20ml/min
	Al plate P = 200W t = 30s
6.5 Resist strip	Ethanol 1+2+3 in Ultrasound bath at roomtemp
	5min, 5min, 2.5min, N_2 blow-dry directly after bath 3

7. Capping layer (also used as LC orientation layer for TN-LCDs)

7.1 Spin-coating	Material: PI Sunever 130 (Nissan Chemical)
	Solvent: 70 % 2-Butoxyethanol, 30 % N-Methyl-2-pyrrolidone
	Mixture PI:Solvent = 2:1
	Max speed 3000 rpm
	Cleaning of fp/bp contacts with Acetone Q-tip directly after deposition
	Hotplate: T = 90°C t = 90s
7.2 Hard bake	Oven: 15min @ 80°C, 75min @ 170°C

A.3. Polymer dispersed liquid crystal (PDLC) display fabrication

Once the TCF is defined as described in appendix A.2, PDLCDs can be realized as follows.

1. Spacer ball spray (frontplane only)

| 1.1 Spray | 1wt% Sekisui SP214 adhesive spacers in IPA |
| | Airbrush spray in dedicated box, 2ml per 2 substrates |

2. Conductive cell frame (backplane only)

2.1 Glue dispensing	Program: as_ph_uni-lfb_PDLC
	Norland NOA68 with 3wt% conductive spacers Sekisui AU215
	Dispense tip: 0.01" Pressure: 3bar Temperature: 25°C
	Distance Tip-Substrate: 50 μm

3. Display cell assembly

3.1 Mask aligner	Frontplane as substrate, backplane as mask
	Once aligned de-activate mask vacuum
	Flood exposure: 2min
	Shield optically active part from UV-radiation
3.2 Define fill-reservoir	Create reservoir around fill opening using UV glue
3.3 Glue curing	10min under UV, shield optically active part from UV-radiation

4. PDLC filling

4.1 Filling	Hot plate: 30°C Vacuum 200mbar
	Place droplet of PDLC mixture at cell opening
4.2 Photo-induced	5min @ 43 mW/cm² in UV chamber
phase separation	

A.4. Twisted nematic (TN) liquid crystal display fabrication

Once the TCF is defined as described in appendix A.2, TN-LCDs can be realized as follows.

1. LC orientation layer
1.1 Rubbing of PI   Rubbing of dummy plate before each substrate
                                Table speed: 600mm/min
                                  Roller speed: 800mm/min
                                  Rub pressure: 0.01mm

2. Spacer ball spray (frontplane only)
2.1 Spray   0.5wt% Sekisui EXH in IPA:H_2O=7:3
                                  Airbrush spray in dedicated box, 2ml per 2 substrates

3. Conductive cell frame (backplane only)
3.1 Glue dispensing   Program: as_ph_uni-lfb_LC
                                  Norland NOA68 with 3wt% conductive spacers Sekisui AU205
                                  Dispense tip: 0.01"   Pressure: 3bar   Temperature: 25°C
                                  Distance Tip-Substrate: 50 µm

4. Display cell assembly
4.1 Mask aligner   Frontplane as substrate, backplane as mask
                                  Once aligned de-activate mask vacuum
                                  Flood exposure: 2min
                                  Shield optically active part from UV-radiation
4.2 Define fill-reservoir   Create reservoir around fill opening using UV glue
4.3 Glue curing   10min under UV, shield optically active part from UV-radiation

5. Liquid crystal filling
5.1 Vaccum filling   pump to 5×10^{-2} mbar   add LC droplet and vent
                                  Add LC mixture to reservoir when necessary
                                  Fill duration: 30-60min
5.2 Sealing   Clean reservoir and add UV glue
                                  10min under UV, shield active-matix from UV-radiation
5.3 Annealing   Heat display above clearing point for 15-30min, cool down slowly

6. Polarizer lamination
6.1 Lamination   90° crossed orientation, parallel to rubbing direction

A.5. a-Si:H AMLCD with CNN pixel electrodes

A.5.1. AMLCD backplane

1. Glass substrates 100mm x 120mm, Label scribing on backside
2. Washing of substrates
3. Sputter deposition of gate metal (Inline sputter tool)
3.1 Gate metal   Material: MoTa   Process: DC
                               Sputter etch   Power: 20 %   t = 4min
                               Program: 3   thickness=150nm

4. Photolithography gate metal
4.1 Resist coating   Resist: LC100
                               Max speed   800 rpm
                               Hotplate: T = 115°C   t = 90s
4.2 Exposure   t = 4s   see product sheet for dosage
                               Mask: Gate, hard contact
4.3 Development   Developer: TMA238WA
                               t = ca. 35s   1min rinsing in DI high-flow
4.4 Wet etching   Honeywell Al etchant @ 40 °C
                               Sputter etch   10%, 2min
                               Duration ca. 15-20s   1min rinsing in DI high-flow
4.5 Resist strip   PRS 1 + 2, Highflow, Hot-DI 1 + 2, Highflow
                               10min each, N_2 blow-dry

5. PECVD layer stack

5.1 Drying	15min @ 120°C
5.2 Pre-conditioning	H_2 plasma
PECVD	O_2 plasma
	SiO_2 dummy deposition
5.3 PECVD	Recipe: TFT-B
	350nm Si_3N_4, 30nm a–Si LDR
	110nm a–Si HDR, 50nm n^+a–Si
5.4 Scratching	Contacts frame gate - source/drain
	Test-TFTs gate contact

6. Sputter deposition source/drain metal (immediately after PECVD)

6.1 Source/drain metal	Material: MoTa	Process: DC
	Sputter etch	Power: 20 % t = 2min
	Program: 3	thickness=200nm

7. Photolithography gate metal

7.1 Resist coating	Resist: LC100	
	Max speed	800 rpm
	Hotplate:	T = 115°C t = 90s
7.2 Exposure	t = 4s	see product sheet for dosage
	Mask: Source/drain, hard contact	
7.3 Development	Developer: TMA238WA	
	t = ca. 35s	1min rinsing in DI high-flow
7.4 Wet etching	Honeywell Al etchant @ 40 °C	
	Sputter etch	10%, 2min
	Duration ca. 15-20s	1min rinsing in DI high-flow
7.5 Resist strip	Butylacetate 1 + 2, IPA, DI high-flow	
	10min each, N_2 blow-dry	

8. Back-channel-etch n^+a–Si

8.1 Pre-conditioning	0,06mbar, Ar 10ml/min, O_2 20ml/min, 400W, Al plate, 10min
plasma etcher	0,03mbar, O_2 10ml/min, CF_4 50ml/min, 100W, Al plate, 1min
8.2 Plasma etching	0,03mbar, O_2 10ml/min, CF_4 50ml/min, 100W, Al plate
	Etch depth: 100nm, etch rate: 50nm/min, t=2min

9. Patterning a–Si

9.1 Resist coating	Resist: LC100	
	Max speed	800 rpm
	Hotplate:	T = 115°C t = 90s
9.2 Exposure	t = 4s	see product sheet for dosage
	Mask: a-Si, hard contact	
9.3 Development	Developer: TMA238WA	
	t = ca. 35s	1min rinsing in DI high-flow
9.4 Annealing resist	30min @ 120°C	
9.5 Pre-conditioning	0,06mbar, Ar 10ml/min, O_2 20ml/min, 400W, Al plate, 10min	
plasma etcher	0,03mbar, O_2 10ml/min, CF_4 50ml/min, 100W, Al plate, 1min	
9.6 Plasma etching	0,03mbar, O_2 10ml/min, CF_4 50ml/min, 100W, Al plate	
	Etch rates: a–Si 50nm/min, Si_3N_4 350nm/min	
9.7 Resist strip	Butylacetate 1 + 2, IPA, DI high-flow	
	10min each, N_2 blow-dry	

10. PECVD semiconductor passivation

10.1 Drying	15min @ 120°C
10.2 PECVD	Recipe: HB_N2O_SiN_250C
	350nm Si_3N_4

11. Patterning of passivation layer and gate dielectric

11.1 Resist coating	Resist: LC100	
	Max speed	800 rpm
	Hotplate:	T = 115°C t = 90s
11.2 Exposure	t = 4s	see product sheet for dosage
	Mask: Via, hard contact	
11.3 Development	Developer: TMA238WA	
	t = ca. 35s	1min rinsing in DI high-flow
11.4 Annealing resist	30min @ 120°C	
11.5 Pre-conditioning	0,06mbar, Ar 10ml/min, O_2 20ml/min, 400W, Al plate, 10min	

plasma etcher	0,05mbar, O_2 10ml/min, CF_4 50ml/min, 200W, Al plate, 2min
11.6 Plasma etching	0,05mbar, O_2 10ml/min, CF_4 50ml/min, 200W, Al plate
	Etch rate: Si_3N_4 350nm/min, t=2min
11.7 Resist strip	Butylacetate 1 + 2, IPA, DI high-flow
	10min each, N_2 blow-dry
11.8 Annealing	150-180°C over night
backplane	for better TFT performance

12. Deposition SWNTs

12.1 Sputter etch	20%, 2min
12.2 SAM deposition	Material: 3-Amino-propyl-triethoxy-silane
	2ml in 200ml H_2O, immersion during 1h
	1min rinsing in DI high-flow
12.3 Spray-coating	HiPCO p suspension
	6ml on each 100mm x 120mm substrate
12.4 Rinsing of	1. Beaker with H_2O at RT t = 10min
surfactant	2. Di high-flow t = 10min
	3. N_2 blow-dry + oven dry 15min @ 120°C

13. Patterning of SWNT layer

13.1 Resist coating	Resist: LC100	
	Max speed	800 rpm
	Hotplate:	T = 115°C t = 90s
13.2 Exposure	t = 4s	see product sheet for dosage
	Mask: transparent conductive film, hard contact	
13.3 Development	Developer: TMA238WA	
	t = ca. 35s	1min rinsing in DI high-flow
13.4 Annealing resist	30min @ 120°C	
13.5 Plasma etching	0,06mbar, Ar 10ml/min, O_2 20ml/min, 200W, Al plate, 1.5min	
	Etch rates: a−Si 50nm/min, Si_3N_4 350nm/min	
13.6 Resist strip	Butylacetate 1 + 2, IPA, DI high-flow	
	10min each, N_2 blow-dry	
13.7 Drying	Oven: T = 120°C t = 15min	

A.5.2. AMLCD frontplane

1. Glass substrates 100mm x 120mm, Label scribing on backside
2. Washing of substrates
3. Sputter deposition of black matrix

3.1 Sputter deposition	Material: MoOxNx−MoTa	Process: DC	
	Sputter etch	Power: 20 % t = 2min	
	Program: 11		
	MoOxNx	thickness = 50nm	v = 0.4m/min
	MoTa	thickness = 120nm	v = 0.58m/min

4. Photolithography black matrix

4.1 Resist coating	Resist: LC100	
	Max speed:	800 rpm
	Hotplate:	T = 115°C t = 90s
4.2 Exposure	t = 4s	see product sheet for dosage
	Mask: Black matrix, soft contact	
4.3 Development	Developer: TMA238WA	
	t = ca. 35s	1min rinsing in DI high-flow
4.4 Wet etching	Mo etchant @ 42 °C	
	Sputter etch	10%, 2min
	Duration ca. 50s	1min rinsing in DI high-flow
4.5 Resist strip	Acetone1, Acetone2, IPA in Ultrasound bath at roomtemp	
	10min, 10min, 5min, N_2 blow-dry directly after IPA	

5. Green color filter

5.1 Resist coating	Resist: Arch SG 2000L	thickness = 0.8 µm	
	Max speed:	600 rpm	t = 40s
	Hotplate:	T = 122°C	t = 90s
5.2 Exposure	t = 28s	see product sheet for dosage	
	Mask: green color filter, soft contact		

5.3 Development	Developer: CD2040		
	t = 18s	1min rinsing in DI high-flow	
5.4 Hard bake	Oven:	T = 220°C	t = 30min
6. Red color filter			
6.1 Resist coating	Resist: Arch CR-7001	thickness = 1.1 μm	
	Max speed:	750 rpm	t = 40s
	Hotplate:	T = 122°C	t = 90s
6.2 Exposure	t = 17s	see product sheet for dosage	
	Mask: red color filter, soft contact		
6.3 Development	Developer: CD2040		
	t = 20s	1min rinsing in DI high-flow	
6.4 Hard bake	Oven:	T = 220°C	t = 30min
7. Blue color filter			
7.1 Resist coating	Resist: Arch SB 2000L	thickness = 0.8 μm	
	Max speed:	600 rpm	t = 40s
	Hotplate:	T = 122°C	t = 90s
7.2 Exposure	t = 30s	see product sheet for dosage	
	Mask: blue color filter, soft contact		
7.3 Development	Developer: CD2040		
	t = 10s	1min rinsing in DI high-flow	
7.4 Hard bake	Oven:	T = 220°C	t = 60min
8. Planarization layer			
8.1 Resist coating	Resist: PC403	thickness = 2.8 μm	
	Max speed:	650 rpm	t = 40s
	Hotplate:	T = 76°C	t = 120s
8.2 Exposure	t = 60s	see product sheet for dosage	
	Mask: Top coat, soft contact		
8.3 Development	Developer: AZ 726 MIF 1:6 with H_2O		
	t = 60s	1min rinsing in DI high-flow	
8.4 Flood exposure	t = 45s		
8.5 Hard bake	Oven:	T = 220°C	t = 60min
9. Deposition SWNTs			
9.1 Sputter etch	20%, 2min		
9.2 SAM deposition	Material: 3-Amino-propyl-triethoxy-silane		
	2ml in 200ml H_2O, immersion during 1h		
	1min rinsing in DI high-flow		
9.3 Spray-coating	HiPCO p suspension		
	6ml on each 10mm x 12mm substrate		
9.4 Rinsing of	1. Beaker with H_2O at RT		t = 10min
surfactant	2. Di high-flow		t = 10min
	3. N_2 blow-dry + oven dry 15min @ 120°C		
9.5 Drying	Oven:	T = 120°C	t = 15min

A.5.3. AMLCD assembly

1. LC orientation layer (front- & backplane)

1.1 Spin-coating	Material: PI Sunever 130 (Nissan Chemical)	
	Solvent: 70 % 2-Butoxyethanol, 30 % N-Methyl-2-pyrrolidone	
	Mixture PI:Solvent = 2:1	
	Max speed	3000 rpm
	Cleaning of fp/bp contacts with Acetone Q-tip directly after deposition	
	Hotplate:	T = 90°C t = 90s
1.2 Hard bake	Oven:	15min @ 80°C, 60min @ 200°C
1.2 Rubbing of PI	Rubbing of dummy plate before each substrate	
	Table speed:	600mm/min
	Roller speed:	800mm/min
	Rub pressure:	0.01mm

2. Spacer ball spray (frontplane only)

| 2.1 Spray | Spacer volume: 25 holes | Speed: 2rpm | F.S.R: -3000V |
| | Nozzle Time: 5s | Ionizer Time: 60s | |

3. Cell frame (backplane only)

3.1 Glue dispensing	Program: AS_10x12_a-Si-Display
	Dispense tip: 0.006" Pressure: 1.8bar Temperature: 45°C
	Distance Tip-Substrate: 50 μm

4. Display cell assembly

4.1 Through contact	Ag glue on 4 contact pads (small amount)
4.2 Mask aligner	Frontplane as substrate, backplane as mask
	Once aligned de-activate mask vacuum
	Flood exposure: 2min
	Shield active-matix from UV-radiation
4.3 Define fill-reservoir	Create reservoir around fill opening using UV glue
4.4 Glue curing	10min under UV, shield active-matix from UV-radiation

4. Liquid crystal filling

4.1 Vaccum filling	30min pump-down add LC droplet and vent
	Add LC mixture to reservoir when necessary
	Fill duration: 30-60min
4.2 Sealing	Clean reservoir and add UV glue
	10min under UV, shield active-matix from UV-radiation
4.3 Annealing	Heat display above clearing point for 15-30min, cool down slowly

5. Driver bonding

5.1 Gate driver	Manual bond tool Parameters according to ACF data sheet
5.2 Source drivers	Semiautomatic bond tool Parameters according to ACF data sheet
	Temperatures to be measured in ACF with thermo-couple

6. Polarizer lamination

6.1 Lamination	90° crossed orientation, parallel to rubbing direction

A.6. TFT process

This process flow can be applied to both glass and plastic substrates.

1. Substrates 50mm x 50mm, label scribing on backside
2. Washing in case of glass substrates
3. Sputter deposition of gate metal (Inline sputter tool)

3.1 Carrier	Material: Al	Process: DC		
pre-conditioning	Sputter etch	Power: 10 %	t = 2min	
		Oszillations: 3	v = 0.57m/min	
3.2 Etch-stop layer	Material: Ta_2O_5	Process: HF		
	Sputter etch	Power: 10 %	t = 2min	
	Program: 9	Oszillations: 7	v = 0.4m/min	
3.3 Gate metal	Material: Al(Nd)	Process: DC		
	Sputter etch	Power: 10 %	t = 2min	
	Program: 5	thickness=150nm		

4. Photolithography gate metal

4.1 Resist coating	Resist: AZ 650-TFP		
	Max speed	2500 rpm	
	Hotplate:	T = 120°C	t = 90s
4.2 Exposure	t = 3.8s	see product sheet for dosage	
	Mask: CNT-TFT REV 2.1 Gate		
4.3 Development	Developer: TMA238WA 7:1 H_2O (ca. 200ml)		
	t = ca. 20s	1min rinsing in DI high-flow	
4.4 Wet etching	Honeywell Al etchant @ 40 °C		
	Duration ca. 35s	1min rinsing in DI high-flow	
4.5 Resist strip	Acetone1, Acetone2, IPA in Ultrasound bath at roomtemp		
	10min, 10min, 5min, N_2 blow-dry directly after IPA		

5. Photolithography for vias in gate dielectric

5.1 Resist coating	Resist: AZ 650-TFP		
	Max speed	2500 rpm	
	Hotplate:	T = 120°C	t = 90s

5.2 Exposure   t = 3.8s   see data sheet for dosage
                    Mask: CNT-TFT REV 2.1 Dielectric
5.3 Development   Developer: TMA238WA 7:1 H_2O
                    t = ca. 20s   1min rinsing in DI high-flow

6. Gate dielectric formation

6.1 Anodisation   Material: Al_2O_3
                   Elektrolyte: 30% stabilized H_2O_2, 2,5l
                   buffered with NH_4OH pH=6..8
                   Current density j: 0.35mA/cm2
                   Gate structure area A: 4.8cm2
                   Current I_A = j*A*#substrates
                   Voltage limit $V_{end} = d_{end}/1.73$nm/V
                   Forming after voltage limit is reached: $V = V_{end}$, 90s

6.2 Resist strip   Acetone1, Acetone2, IPA in Ultrasound bath at roomtemp
                   10min, 10min, 5min, N2 blow-dry directly after IPA

7. Deposition of SWNTs

7.1 Sputter etch   Power: 10 %   t = 2min
7.2 SAM deposition   Material: 3-Amino-propyl-triethoxy-silane
                     2ml in 200ml H_2O, immersion during 1h
                     1min rinsing in DI high-flow
7.3 Spin-coating   **Methanol cartridge, 0.01" tip (red)**
of CNTs   **CNT cartridge, 0.006" tip (pink)**
                     CNT pressure 10psi, Methanol pressure to be adapted
                     3000rpm
7.4 Rinsing of   1. Beaker with H_2O at RT   t = 10min
surfactant   2. High-flow   t = 10min
                     3. N_2 blow-dry

8. Source/Drain metal

8.1 Resist coating   Resist: ma-N 1420, Syringe with 1μm filter
                     Max speed   2600 rpm
                     Hotplate: T = 120°C   t = 120s
8.2 Exposure   t = 70s   see data sheet for dosage
                     Maske: CNT-TFT REV 2.1 Source/Drain
8.3 Developement   Developer: ma-D 533s
                     t = 120s   1min rinsing in high-flow
8.4 Metal evap.   Material: Pd
                     Thickness: 300nm
8.5 Liftoff   Acetone1, Acetone2, IPA in ultrasound bath at roomtemp
                     only short bursts of ultrasound
                     Acetone1 to be renewed after each substrate

9. Patterning of SWNT layer

9.1 Resist coating   Resist: AZ 650-TFP
                     Max speed   2500 rpm
                     Hotplate: T = 120°C   t = 90s
9.2 Exposure   t = 3.8s   see product sheet for dosage
                     Mask: CNT-TFT REV 2.1 Gate
9.3 Development   Developer: TMA238WA 7:1 H2O (ca. 200ml)
                     t = ca. 20s   1min rinsing in DI high-flow
9.4 RIE   p: 0.06mbar   Ar 10ml/min, O_2 20ml/min
                     Al plate   P = 200W   t = 30s
9.5 Resist strip   Acetone1, Acetone2, IPA at roomtemp
                     only short bursts of US
                     10min, 10min, 5min, N2 blow-dry directly after IPA

B. Roughness value definitions

The result of an AFM scan are height values for each of the evaluated coordinates $Z(x,y)$. From these different roughness values can be calculated. The following measures for surface roughness are used in this work.

The profile height S_y gives the peak-to-valley height.

$$S_y = Z_{max} - Z_{min} \tag{B.1}$$

The ten point height S_z represents the profile height averaged over the 5 highest and lowest points.

$$S_z = \frac{Z_{max1} + Z_{max2} + Z_{max3} + Z_{max4} + Z_{max5} - Z_{min1} - Z_{min2} - Z_{min3} - Z_{min4} - Z_{min5}}{5} \tag{B.2}$$

The roughness average S_a is defined by

$$S_a = \frac{1}{N} \sum_{i=1}^{N} |Z_i - Z_m| \tag{B.3}$$

where N is the number of measured points and

$$Z_m = \frac{1}{N} \sum_{i=1}^{N} |Z_i| \tag{B.4}$$

.

C. Further Monte-Carlo simulation results

Table C.1.: Percolation data for fig. 3.5.

M	$p_{0.5\text{mix}}$	$p_{50\text{mix}}$	$p_{99.5\text{mix}}$	$p_{0.5\text{m}}$	$p_{50\text{m}}$	$p_{99.5\text{m}}$	$\frac{p_{50\text{m}}}{p_{50\text{mix}}}$	$\frac{TZ_m}{TZ_{mix}}$	SR	t_M [s]
10	2.01	2.92	4.04	5.77	8.77	12.46	3.0	3.3	1.7	278
15	2.43	3.48	4.80	7.10	10.45	14.55	3.0	3.1	2.3	385
25	2.84	4.12	5.54	8.64	12.31	17.04	3.0	3.1	3.1	624
50	3.45	4.75	6.36	10.03	14.21	19.38	3.0	3.2	3.7	1547
100	3.69	5.15	6.83	11.03	15.48	20.82	3.0	3.1	4.2	4742
200	3.91	5.38	7.10	11.70	16.13	21.83	3.0	3.2	4.6	16285
300	3.96	5.45	7.32	11.83	16.41	21.99	3.0	3.0	4.5	35735
400	4.10	5.52	7.39	11.86	16.51	22.18	3.0	3.1	4.5	61704
1000	4.12	5.59	7.36	12.12	16.81	22.30	3.0	3.1	4.8	359236

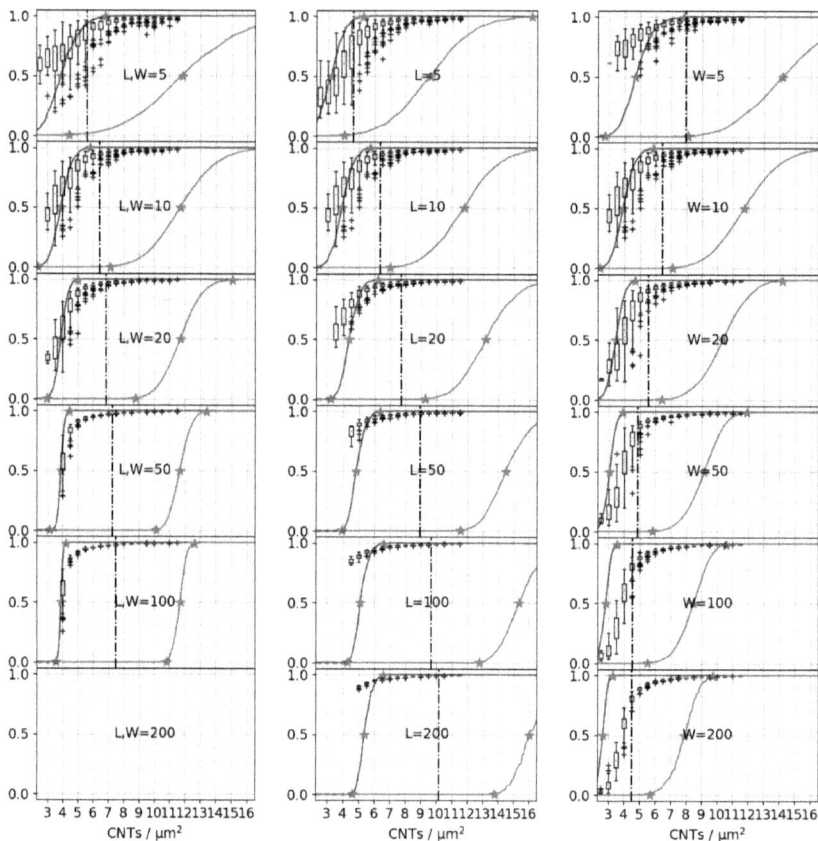

(a) Square scaling, $L = W$; no simulation   (b) Length variation, $W = 10\,\mu m$ (c) Width variation, $L = 10\,\mu m$
data for $L,W = 200\,\mu m$

Figure C.1.: Percolation probability of mix-CNN (blue) and met-CNN (red) for CNT length distribution as shown in fig. 3.11 and different channel geometries; stars indicate positions of $p_{0.5}$, p_{50} and $p_{99.5}$; vertical dashed line is center of SR; box and whisker plot shows ratio of electrically active vs. deposited CNTs from 100 random CNNs per density (red line: median, box: upper and lower quartille, whiskers: upper and lower limit, pluses: outliers).

D. Further CNT-TFT results

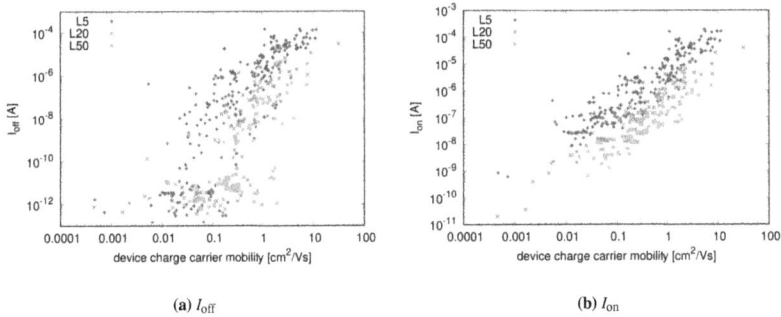

(a) I_{off}

(b) I_{on}

Figure D.1.: I_{off} and I_{on} vs. μ_{device} of a range of mix-CNN TFTs on glass and plastic, $V_{DS} = -0.1$ V; same devices as in fig. 7.7.

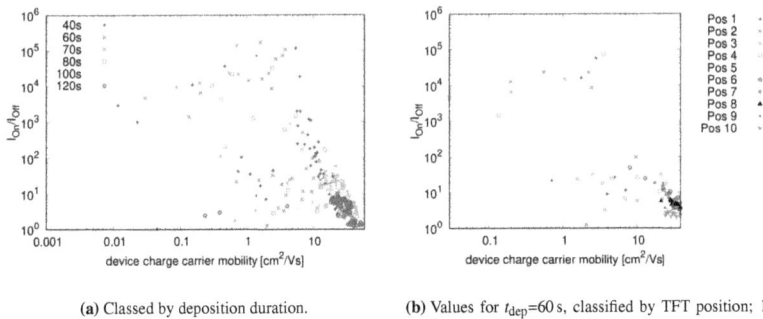

(a) Classed by deposition duration.

(b) Values for t_{dep}=60 s, classified by TFT position; 1-10 from outside to center of substrate.

Figure D.2.: On/off-ratio vs. μ_{device} for TFTs fabricated with 98 % semiconducting nanotubes and static dispense head, $V_{DS} = -0.1$ V.

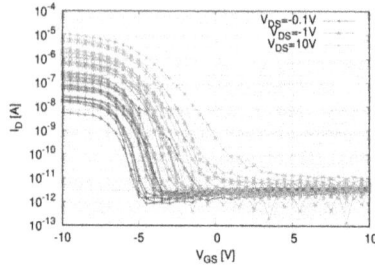

Figure D.3.: Transfer characteristics for 10 TFTs with 10 s deposition duration and $L_C = W_C = 50\,\mu m$ for different V_{DS}; 98 % semiconducting nanotubes and moving dispense head.

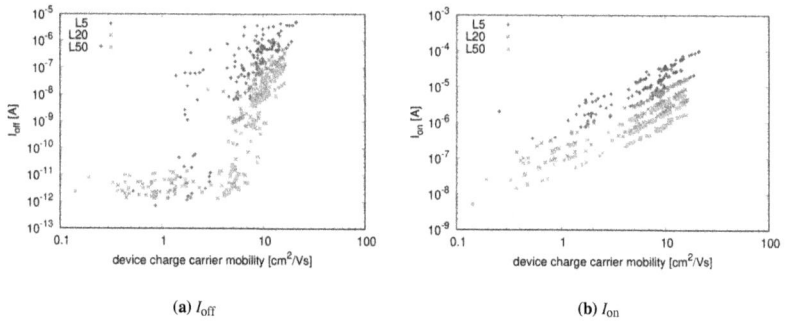

(a) I_{off}

(b) I_{on}

Figure D.4.: I_{off} and I_{on} vs. μ_{device} for TFTs fabricated with 98 % semiconducting nanotubes and moving dispense head, $V_{DS} = -0.1\,V$; same devices and deposition times as in fig. 7.8.

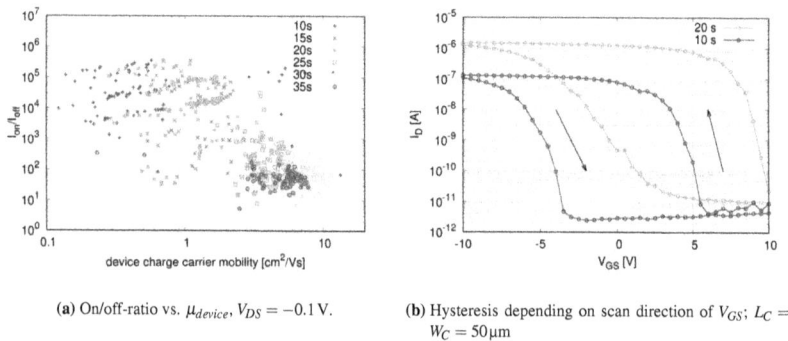

(a) On/off-ratio vs. μ_{device}, $V_{DS} = -0.1\,V$.

(b) Hysteresis depending on scan direction of V_{GS}; $L_C = W_C = 50\mu m$

Figure D.5.: Performance of Au S/D bottom contact TFTs fabricated with 98 % semiconducting nanotubes and moving dispense head.

Bibliography

[1] ABDALLA, S. ; AL-MARZOUKI, F. ; AL-GHAMDI, Ahmed A. ; ABDEL-DAIEM, A.: Different Technical Applications of Carbon Nanotubes. In: *Nanoscale Research Letters* 10 (2015), September. http://dx.doi.org/10.1186/s11671-015-1056-3. – DOI 10.1186/s11671–015–1056–3. – ISSN 1931–7573

[2] ALFONSI, Jessica: *Electronic Band Structure of a Single-Walled Carbon Nanotube by the Zone-Folding Method.* http://demonstrations.wolfram.com/ ElectronicBandStructureOfASingleWalledCarbonNanotubeByTheZon/. Version: March 2011

[3] ANISIMOV, Anton S. ; BROWN, David P. ; MIKLADAL, Bjorn F. ; SUILLEABHAIN, Liam O. ; PARIKH, Kunjal ; SOININEN, Erkki ; SONNINEN, Martti ; TIAN, Dewei ; VARJOS, Ilkka ; VUOHE-LAINEN, Risto: 16.3: Printed Touch Sensors Using Carbon NanoBud® Material. In: *SID Symposium Digest of Technical Papers* 45 (2014), June, Nr. 1, 200–203. http://dx.doi.org/10.1002/ j.2168-0159.2014.tb00055.x. – DOI 10.1002/j.2168–0159.2014.tb00055.x. – ISSN 2168–0159

[4] ARNOLD, K. ; HENNRICH, F. ; KRUPKE, R. ; LEBEDKIN, S. ; KAPPES, M. M.: Length separation studies of single walled carbon nanotube dispersions. In: *physica status solidi (b)* 243 (2006), Nr. 13, 3073–3076. http://dx.doi.org/10.1002/pssb.200669196. – DOI 10.1002/pssb.200669196

[5] ARNOLD, M. S. ; STUPP, S. I. ; HERSAM, M. C.: Enrichment of single-walled carbon nanotubes by diameter in density gradients. In: *Nano Lett* 5 (2005), Nr. 4, S. 713–718

[6] ARNOLD, Michael S. ; GREEN, Alexander A. ; HULVAT, James F. ; STUPP, Samuel I. ; HERSAM, Mark C.: Sorting carbon nanotubes by electronic structure using density differentiation. In: *Nature Nanotechnology* 1 (2006), Nr. 1, S. 60–65

[7] BACHILO, Sergei M. ; BALZANO, Leandro ; HERRERA, Jose E. ; POMPEO, Francisco ; RESASCO, Daniel E. ; WEISMAN, R. B.: Narrow (n,m)-Distribution of Single-Walled Carbon Nanotubes Grown Using a Solid Supported Catalyst. In: *Journal of the American Chemical Society* 125 (2003), Nr. 37, 11186–11187. http://dx.doi.org/10.1021/ja036622c. – DOI 10.1021/ja036622c

[8] BALASUBRAMANIAN, Kannan ; SORDAN, Roman ; BURGHARD, Marko ; KERN, Klaus: A Selective Electrochemical Approach to Carbon Nanotube Field-Effect Transistors. In: *Nano Letters* 4 (2004), May, Nr. 5, 827–830. http://dx.doi.org/10.1021/nl049806d. – DOI 10.1021/nl049806d

[9] BERRY, Robert W. ; HALL, Peter M. ; HARRIS, Murray T.: *Thin Film Technology.* New York : Litton Educational Publishing, Inc., 1968

[10] BETHUNE, D. S. ; KIANG, C. H. ; VRIES, M. S. d. ; GORMAN, G. ; SAVOY, R. ; VAZQUEZ, J. ; BEYERS, R.: Cobalt-catalysed growth of carbon nanotubes with single-atomic-layer walls. In: *Nature* 363 (1993), June, Nr. 6430, 605–607. http://dx.doi.org/10.1038/363605a0. – DOI 10.1038/363605a0. – ISSN 1476–4687

[11] BHUSHAN, Bharat ; FUCHS, Harald ; HOSAKA, Sumio: *Applied Scanning Probe Methods.* 2004

[12] BLACKBURN, Jeffrey L. ; BARNES, Teresa M. ; BEARD, Matthew C. ; KIM, Yong-Hyun ; TENENT, Robert C. ; MCDONALD, Timothy J. ; TO, Bobby ; COUTTS, Timothy J. ; HEBEN, Michael J.: Transparent Conductive Single-Walled Carbon Nanotube Networks with Precisely Tunable Ratios of

Semiconducting and Metallic Nanotubes. In: *ACS Nano* 2 (2008), June, Nr. 6, 1266–1274. http://dx.doi.org/10.1021/nn800200d. – DOI 10.1021/nn800200d

[13] BRADY, Gerald J. ; JOO, Yongho ; WU, Meng-Yin ; SHEA, Matthew J. ; GOPALAN, Padma ; ARNOLD, Michael S.: Polyfluorene-Sorted, Carbon Nanotube Array Field-Effect Transistors with Increased Current Density and High On/Off Ratio. In: *ACS Nano* 8 (2014), Nr. 11, 11614–11621. http://dx.doi.org/10.1021/nn5048734. – DOI 10.1021/nn5048734. – ISSN 1936–0851

[14] BRESENHAM, J. E.: Algorithm for Computer Control of a Digital Plotter. In: *IBM Syst. J.* 4 (1965), March, Nr. 1, 25–30. http://dx.doi.org/10.1147/sj.41.0025. – DOI 10.1147/sj.41.0025. – ISSN 0018–8670

[15] BRONIKOWSKI, Michael J. ; WILLIS, Peter A. ; COLBERT, Daniel T. ; SMITH, K.A. ; SMALLEY, Richard E.: Gas-phase production of carbon single-walled nanotubes from carbon monoxide via the HiPco process: A parametric study. In: *Journal of Vacuum Science & Technology A* 19 (2001), Nr. 4, S. 1800–1805

[16] BROTHERTON, S. D.: *Introduction to Thin Film Transistors: Physics and Technology of TFTs.* Springer International Publishing, 2013 https://www.springer.com/de/book/9783319000015. – ISBN 978–3–319–00001–5

[17] CASTELLANO, Joseph A.: *Liquid gold: the story of liquid crystal displays and the creation of an industry.* Singapore; Hackensack, NJ; London : World Scientific, 2005 http://public.eblib.com/choice/publicfullrecord.aspx?p=239632. – ISBN 978–981–256–584–6 978–981–238–956–5 978–1–281–87721–5. – OCLC: 232157254

[18] CHEN, Janglin ; CRANTON, Wayne ; FIHN, Mark: *Handbook of visual display technology.* Springer, 2012 http://www.zigbee-devcon-europe.de/fileadmin/user_upload/pdf/ed2011/day1/Handbook/Handbook_of_of_Visual_Display_Technology.pdf

[19] CHEN, Janglin (eds.) ; CRANTON, Wayne (eds.) ; FIHN, Mark (eds.): *Handbook of Visual Display Technology.* 2. Springer International Publishing, 2016 https://www.springer.com/de/book/9783319143453. – ISBN 978–3–319–14345–3

[20] COLLINS, Philip G. ; ARNOLD, Michael S. ; AVOURIS, Phaedon: Engineering Carbon Nanotubes and Nanotube Circuits Using Electrical Breakdown. In: *Science* 292 (2001), April, Nr. 5517, 706–709. http://dx.doi.org/10.1126/science.1058782. – DOI 10.1126/science.1058782

[21] COLLINS, Philip G. ; BRADLEY, Keith ; ISHIGAMI, Masa ; ZETTL, A.: Extreme Oxygen Sensitivity of Electronic Properties of Carbon Nanotubes. In: *Science* 287 (2000), March, Nr. 5459, 1801–1804. http://dx.doi.org/10.1126/science.287.5459.1801. – DOI 10.1126/science.287.5459.1801

[22] COMMUNITY, Scipy: *numpy.random.beta — NumPy v1.16 Manual.* https://docs.scipy.org/doc/numpy/reference/generated/numpy.random.beta.html. Version: January 2019

[23] CUI, Kehang ; ANISIMOV, Anton S. ; CHIBA, Takaaki ; FUJII, Shunjiro ; KATAURA, Hiromichi ; NASIBULIN, Albert G. ; CHIASHI, Shohei ; KAUPPINEN, Esko I. ; MARUYAMA, Shigeo: Air-stable high-efficiency solar cells with dry-transferred single-walled carbon nanotube films. In: *Journal of Materials Chemistry A* 2 (2014), July, Nr. 29, 11311–11318. http://dx.doi.org/10.1039/C4TA01353K. – DOI 10.1039/C4TA01353K. – ISSN 2050–7496

[24] DILLON, A. C. ; GENNETT, T. ; JONES, K. M. ; ALLEMAN, J. L. ; PARILLA, P. A. ; HEBEN, M. J.: A Simple and Complete Purification of Single-Walled Carbon Nanotube Materials. In: *Advanced Materials* 11 (1999), Nr. 16, 1354–1358. http://dx.doi.org/10.1002/(SICI)1521-4095(199911)11:16<1354::AID-ADMA1354>3.0.CO;2-N. – DOI 10.1002/(SICI)1521–4095(199911)11:16<1354::AID–ADMA1354>3.0.CO;2–N

[25] DIMAKI, Maria ; BOGGILD, Peter: Dielectrophoresis of carbon nanotubes using microelectrodes: a numerical study. In: *Nanotechnology* 15 (2004), August, Nr. 8, 1095–1102. http://dx.doi.org/10.1088/0957-4484/15/8/039. – DOI 10.1088/0957–4484/15/8/039. – ISSN 0957–4484

[26] DRESSELHAUS, M. S. ; DRESSELHAUS, G. ; SAITO, Riichiro: Carbon fibers based on C60 and their symmetry. In: *Physical Review B* 45 (1992), March, Nr. 11, 6234–6242. http://dx.doi.org/10.1103/PhysRevB.45.6234. – DOI 10.1103/PhysRevB.45.6234

[27] DRESSELHAUS, Mildred S. ; DRESSELHAUS, Gene ; AVOURIS, Phaedon: *Carbon Nanotubes - Synthesis, Structure, Properties, and Applications.* Springer, 2001

[28] DURKOP, T. ; GETTY, S. A. ; COBAS, Enrique ; FUHRER, M. S.: Extraordinary Mobility in Semiconducting Carbon Nanotubes. In: *Nano Letters* 4 (2004), January, Nr. 1, 35–39. http://dx.doi.org/10.1021/nl034841q. – DOI 10.1021/nl034841q

[29] EGELHAAF, Jan: *Reflective Flüssigkristall-Farbbildschirme mit hoher optischer Apertur und reduziertem Übersprechen.* Stuttgart, Universität Stuttgart, Dissertation, April 2001

[30] FOUQUET, M. ; BAYER, B. C. ; ESCONJAUREGUI, S. ; BLUME, R. ; WARNER, J. H. ; HOFMANN, S. ; SCHLÖGL, R. ; THOMSEN, C. ; ROBERTSON, J.: Highly chiral-selective growth of single-walled carbon nanotubes with a simple monometallic Co catalyst. In: *Physical Review B* 85 (2012), June, Nr. 23, 235411. http://dx.doi.org/10.1103/PhysRevB.85.235411. – DOI 10.1103/PhysRevB.85.235411

[31] FUHRER, M. S. ; NYGÅRD, J. ; SHIH, L. ; FORERO, M. ; YOON, Young-Gui ; MAZZONI, M. S. C. ; CHOI, Hyoung J. ; IHM, Jisoon ; LOUIE, Steven G. ; ZETTL, A. ; MCEUEN, Paul L.: Crossed Nanotube Junctions. In: *Science* 288 (2000), April, Nr. 5465, 494–497. http://dx.doi.org/10.1126/science.288.5465.494. – DOI 10.1126/science.288.5465.494

[32] GAO, Jing ; WANG, Wen-Yi ; CHEN, Li-Ting ; CUI, Li-Jun ; HU, Xiao-Yan ; GENG, Hong-Zhang: Optimizing processes of dispersant concentration and post-treatments for fabricating single-walled carbon nanotube transparent conducting films. In: *Applied Surface Science* 277 (2013), July, 128–133. http://dx.doi.org/10.1016/j.apsusc.2013.04.012. – DOI 10.1016/j.apsusc.2013.04.012. – ISSN 0169–4332

[33] GAVRYUSHIN, Vladimir: *Graphene Brillouin Zone and Electronic Energy Dispersion.* http://demonstrations.wolfram.com/GrapheneBrillouinZoneAndElectronicEnergyDispersion/. Version: March 2011

[34] GENG, Hong Z. ; KIM, Ki K. ; SO, Kang P. ; LEE, Young S. ; CHANG, Youngkyu ; LEE, Young H.: Effect of Acid Treatment on Carbon Nanotube-Based Flexible Transparent Conducting Films. In: *Journal of the American Chemical Society* 129 (2007), Nr. 25, S. 7758–7759

[35] GHOSH, Saunab ; BACHILO, Sergei M. ; WEISMAN, R. B.: Advanced sorting of single-walled carbon nanotubes by nonlinear density-gradient ultracentrifugation. In: *Nature Nanotechnology* 5 (2010), June, Nr. 6, 443–450. http://dx.doi.org/10.1038/nnano.2010.68. – DOI 10.1038/nnano.2010.68. – ISSN 1748–3387

[36] GOETTLING, Silke ; BRILL, Jochen ; FRUEHAUF, Norbert ; PFLAUM, Jens ; MARGALLO-BALBAS, Eduardo: Pentacene Organic Thin Film Transistors with Anodized Gate Dielectric. In: *Proc. SPIE* vol. 5940, 2005, S. 183–191

[37] GOOCH, C. H. ; TARRY, H. A.: The optical properties of twisted nematic liquid crystal structures with twist angles smaller or equal 90 degrees. In: *Journal of Physics D: Applied Physics* 8 (1975), September, Nr. 13, 1575–1584. http://dx.doi.org/10.1088/0022-3727/8/13/020. – DOI 10.1088/0022–3727/8/13/020. – ISSN 0022–3727

[38] GSALLER, Guenther: *Some Examples of Molecule Orbitals.* http://demonstrations.wolfram.com/SomeExamplesOfMoleculeOrbitals/. Version: October 2013

[39] HAACKE, G.: New figure of merit for transparent conductors. In: *Journal of Applied Physics* 47 (1976), September, Nr. 9, 4086–4089. http://dx.doi.org/10.1063/1.323240. – DOI 10.1063/1.323240. – ISSN 0021–8979, 1089–7550

[40] HARUTYUNYAN, A. R. ; CHEN, G. ; PARONYAN, T. M. ; PIGOS, E. M. ; KUZNETSOV, O. A. ; HEWAPARAKRAMA, K. ; KIM, S. M. ; ZAKHAROV, D. ; STACH, E. A. ; SUMANASEKERA, G. U.: Preferential Growth of Single-Walled Carbon Nanotubes with Metallic Conductivity. In: *Science* 326 (2009), October, Nr. 5949, 116–120. http://dx.doi.org/10.1126/science.1177599. – DOI 10.1126/science.1177599. – ISSN 0036–8075

[41] HASSANIEN, A ; TOKUMOTO, M ; UMEK, P ; VRBANIČ, D ; MOZETIČ, M ; MIHAILOVIĆ, D ; VENTURINI, P: Selective etching of metallic single-wall carbon nanotubes with hydrogen plasma. In: *Nanotechnology* 16 (2005), February, Nr. 2, 278–281. http://dx.doi.org/10.1088/0957-4484/16/2/017. – DOI 10.1088/0957–4484/16/2/017. – ISSN 0957–4484

[42] HE, Maoshuai ; JIANG, Hua ; LIU, Bilu ; FEDOTOV, Pavel V. ; CHERNOV, Alexander I. ; OBRAZTSOVA, Elena D. ; CAVALCA, Filippo ; WAGNER, Jakob B. ; HANSEN, Thomas W. ; ANOSHKIN, Ilya V. ; OBRAZTSOVA, Ekaterina A. ; BELKIN, Alexey V. ; SAIRANEN, Emma ; NASIBULIN, Albert G. ; LEHTONEN, Juha ; KAUPPINEN, Esko I.: Chiral-Selective Growth of Single-Walled Carbon Nanotubes on Lattice-Mismatched Epitaxial Cobalt Nanoparticles. In: *Scientific Reports* 3 (2013), March. http://dx.doi.org/10.1038/srep01460. – DOI 10.1038/srep01460

[43] HECHT, David S. ; HEINTZ, Amy M. ; LEE, Roland ; HU, Liangbing ; MOORE, Bryon ; CUCKSEY, Chad ; RISSER, Steven: High conductivity transparent carbon nanotube films deposited from superacid. In: *Nanotechnology* 22 (2011), February, Nr. 7, 075201. http://dx.doi.org/10.1088/0957-4484/22/7/075201. – DOI 10.1088/0957–4484/22/7/075201. – ISSN 0957–4484

[44] HECHT, David S. ; SIERROS, Konstantinos A. ; LEE, Roland S. ; LADOUS, Corinne ; NIU, Chunming ; BANERJEE, Derrick A. ; CAIRNS, Darran R.: Transparent conductive carbon-nanotube films directly coated onto flexible and rigid polycarbonate. In: *Journal of the Society for Information Display* 19 (2011), February, Nr. 2, 157–162. http://dx.doi.org/10.1889/JSID19.2.157. – DOI 10.1889/JSID19.2.157. – ISSN 1938–3657

[45] HECHT, David S. ; THOMAS, David ; HU, Liangbing ; LADOUS, Corinne ; LAM, Tom ; PARK, Youngbae ; IRVIN, Glen ; DRZAIC, Paul: Carbon-nanotube film on plastic as transparent electrode for resistive touch screens. In: *Journal of the Society for Information Display* 17 (2009), November, Nr. 11, 941–946. http://dx.doi.org/10.1889/JSID17.11.941. – DOI 10.1889/JSID17.11.941. – ISSN 1938–3657

[46] HEERMANN, Dieter W. ; BINDER, K.: *Monte Carlo Simulation in Statistical Physics.* Springer-Verlag Berlin Heidelberg, 2010 http://www.ulb.tu-darmstadt.de/tocs/55667112.pdf

[47] HELFRICH, Wolfgang ; SCHADT, Martin: *Lichtsteuerzelle.* Basel, February 1973

[48] HELLSTROM, Sondra L. ; VOSGUERITCHIAN, Michael ; STOLTENBERG, Randall M. ; IRFAN, Irfan ; HAMMOCK, Mallory ; WANG, Yinchao B. ; JIA, Chuancheng ; GUO, Xuefeng ; GAO, Yongli ; BAO, Zhenan: Strong and Stable Doping of Carbon Nanotubes and Graphene by MoOx for Transparent Electrodes. In: *Nano Letters* 12 (2012), July, Nr. 7, 3574–3580. http://dx.doi.org/10.1021/nl301207e. – DOI 10.1021/nl301207e. – ISSN 1530–6984

[49] HERGERT, Steffen: *Aufbau von OLED-Displays und ihre Verkapselung.* Uelvesbüll : Der Andere Verlag, 2011. – ISBN 978–3–86247–131–7

[50] HERSAM, Mark C.: Progress towards monodisperse single-walled carbon nanotubes. In: *Nature Nanotechnology* 3 (2008), July, Nr. 7, 387–394. http://dx.doi.org/10.1038/nnano.2008.135. – DOI 10.1038/nnano.2008.135. – ISSN 1748–3387

[51] HILL, Christian: *Converting a spectrum to a colour.* https://scipython.com/blog/converting-a-spectrum-to-a-colour/. Version: March 2016

[52] HOU, Peng-Xiang ; LIU, Chang ; CHENG, Hui-Ming: Purification of carbon nanotubes. In: *Carbon* 46 (2008), December, Nr. 15, 2003–2025. http://dx.doi.org/10.1016/j.carbon.2008.09.009. – DOI 10.1016/j.carbon.2008.09.009. – ISSN 0008–6223

[53] HU, L. ; HECHT, D. S. ; GRÜNER, G.: Percolation in Transparent and Conducting Carbon Nanotube Networks. In: *Nano Letters* 4 (2004), December, Nr. 12, 2513–2517. http://dx.doi.org/10.1021/nl048435y. – DOI 10.1021/nl048435y

[54] HUANG, Houjin ; MARUYAMA, Ryuichiro ; NODA, Kazuhiro ; KAJIURA, Hisashi ; KADONO, Koji: Preferential Destruction of Metallic Single-Walled Carbon Nanotubes by Laser Irradiation. In: *The Journal of Physical Chemistry B* 110 (2006), April, Nr. 14, 7316–7320. http://dx.doi.org/10.1021/jp056684k. – DOI 10.1021/jp056684k

[55] HUANG, Limin ; JIA, Zhang ; O'BRIEN, Stephen: Orientated assembly of single-walled carbon nanotubes and applications. In: *Journal of Materials Chemistry* 17 (2007), Nr. 37, 3863–3874. http://dx.doi.org/10.1039/b702080e

[56] HUANG, Xueying ; MCLEAN, Robert S. ; ZHENG, Ming: High-Resolution Length Sorting and Purification of DNA-Wrapped Carbon Nanotubes by Size-Exclusion Chromatography. In: *Analytical Chemistry* 77 (2005), October, Nr. 19, 6225–6228. http://dx.doi.org/10.1021/ac0508954. – DOI 10.1021/ac0508954

[57] IIJIMA, Sumio: Helical microtubules of graphitic carbon. In: *Nature* 354 (1991), November, Nr. 6348, 56. http://dx.doi.org/10.1038/354056a0. – DOI 10.1038/354056a0. – ISSN 1476–4687

[58] IIJIMA, Sumio ; ICHIHASHI, Toshinari: Single-shell carbon nanotubes of 1-nm diameter. In: *Nature* 363 (1993), June, Nr. 6430, 603–605. http://dx.doi.org/10.1038/363603a0. – DOI 10.1038/363603a0. – ISSN 1476–4687

[59] ISEMANN, Heike: *Optimierung der Kohlenstoff Nanoröhren Abscheidung für Dünnschichttransistoren.* Stuttgart, Universität Stuttgart, Studienarbeit, August 2007

[60] JACKSON, Roderick ; GRAHAM, Samuel: Specific contact resistance at metal/carbon nanotube interfaces. In: *Applied Physics Letters* 94 (2009), January, Nr. 1, 012109–3. http://dx.doi.org/10.1063/1.3067819. – DOI 10.1063/1.3067819

[61] JASTI, Ramesh ; BHATTACHARJEE, Joydeep ; NEATON, Jeffrey B. ; BERTOZZI, Carolyn R.: Synthesis, Characterization, and Theory of [9]-, [12]-, and [18]Cycloparaphenylene: Carbon Nanohoop Structures. In: *Journal of American Chemical Society* 130 (2008), Nr. 52, S. 17646–17647

[62] JAVEY, Ali ; KIM, Hyoungsub ; BRINK, Markus ; WANG, Qian ; URAL, Ant ; GUO, Jing ; MCINTYRE, Paul ; MCEUEN, Paul ; LUNDSTROM, Mark ; DAI, Hongjie: High-k dielectrics for advanced carbon-nanotube transistors and logic gates. In: *Nature Materials* 1 (2002), December, Nr. 4, 241–246. http://dx.doi.org/10.1038/nmat769. – DOI 10.1038/nmat769. – ISSN 1476–1122

[63] JELTING, Silke: *Organische Dünnschichttransistoren für den Einsatz in flexiblen Aktiv Matrix Anzeigen.* 1. Aachen : Shaker, 2010. – ISBN 978–3–8322–9391–8

[64] JELTING, Sven: *Konzepte zur Ansteuerung von Matrix adressierten Bildschirmen und einer OLED-Bildpunktschaltung mit Stromrückführung.* 1., Aachen : Shaker, 2009. – ISBN 978–3–8322–8312–4

[65] JORIO, Ado ; DRESSELHAUS, Gene ; DRESSELHAUS, Mildred S.: *Carbon nanotubes - Advanced Topics in the Synthesis, Structure, Properties and Applications.* Springer, 2008 (Topics in Applied Physics 111). – ISBN 3–540–72864–3 978–3–540–72864–1

[66] KANE, C. L. ; MELE, E. J.: Size, Shape, and Low Energy Electronic Structure of Carbon Nanotubes. In: *Physical Review Letters* 78 (1997), March, Nr. 10, 1932–1935. http://dx.doi.org/10.1103/PhysRevLett.78.1932. – DOI 10.1103/PhysRevLett.78.1932

[67] KANG, Lixing ; HU, Yue ; LIU, Lili ; WU, Juanxia ; ZHANG, Shuchen ; ZHAO, Qiuchen ; DING, Feng ; LI, Qingwen ; ZHANG, Jin: Growth of Close-Packed Semiconducting Single-Walled Carbon Nanotube Arrays Using Oxygen-Deficient TiO2 Nanoparticles as Catalysts. In: *Nano Letters* (2014). http://dx.doi.org/10.1021/nl5037325. – DOI 10.1021/nl5037325. – ISSN 1530–6984

[68] KANG, Seong J. ; KOCABAS, Coskun ; OZEL, Taner ; SHIM, Moonsub ; PIMPARKAR, Ninad ; ALAM, Muhammad A. ; ROTKIN, Slava V. ; ROGERS, John A.: High-performance electronics using dense, perfectly aligned arrays of single-walled carbon nanotubes. In: *Nature Nanotechnology* 2 (2007), April, Nr. 4, 230–236. http://dx.doi.org/10.1038/nnano.2007.77. – DOI 10.1038/nnano.2007.77. – ISSN 1748–3387

[69] KASKELA, Antti ; NASIBULIN, Albert G. ; TIMMERMANS, Marina Y. ; AITCHISON, Brad ; PAPADIMITRATOS, Alexios ; TIAN, Ying ; ZHU, Zhen ; JIANG, Hua ; BROWN, David P. ; ZAKHIDOV, Anvar ; KAUPPINEN, Esko I.: Aerosol-Synthesized SWCNT Networks with Tunable Conductivity and Transparency by a Dry Transfer Technique. In: *Nano Letters* 10 (2010), November, Nr. 11, 4349–4355. http://dx.doi.org/10.1021/nl101680s. – DOI 10.1021/nl101680s. – ISSN 1530–6984

[70] KIM, Woong ; JAVEY, Ali ; VERMESH, Ophir ; WANG, Qian ; YIMING, Li ; DAI, Hongjie: Hysteresis caused by water molecules in carbon nanotube field-effect transistors. In: *Nano Letters* 3 (2003), Nr. 2, S. 193–198

[71] KLEINER, Alex ; EGGERT, Sebastian: Curvature, hybridization, and STM images of carbon nanotubes. In: *Physical Review B* 64 (2001), August, Nr. 11, 113402. http://dx.doi.org/10.1103/PhysRevB.64.113402. – DOI 10.1103/PhysRevB.64.113402

[72] KNORR, Sebastian: *Dünnschichttransistoren mit rein halbleitenden Kohlenstoff Nanoröhren.* Stuttgart, Universität Stuttgart, Studienarbeit, September 2010

[73] KOCABAS, C. ; PIMPARKAR, N. ; YESILYURT, O. ; KANG, S. J. ; ALAM, M. A. ; ROGERS, J. A.: Experimental and Theoretical Studies of Transport through Large Scale, Partially Aligned Arrays of Single-Walled Carbon Nanotubes in Thin Film Type Transistors. In: *Nano Letters* 7 (2007), May, Nr. 5, 1195–1202. http://dx.doi.org/10.1021/nl062907m. – DOI 10.1021/nl062907m

[74] KROTO, H. W. ; HEATH, J. R. ; O'BRIEN, S. C. ; CURL, R. F. ; SMALLEY, Richard E.: C60: Buckminsterfullerene. In: *Nature* 318 (1985), S. 162–163

[75] KRÄTSCHMER, W. ; LAMB, Lowell D. ; FOSTIROPOULOS, K. ; HUFFMAN, Donald R.: Solid C 60 : a new form of carbon. In: *Nature* 347 (1990), September, Nr. 6291, 354–358. http://dx.doi.org/10.1038/347354a0. – DOI 10.1038/347354a0. – ISSN 1476–4687

[76] KRUPKE, Ralph ; HENNRICH, Frank ; KAPPES, Manfred M. ; LOHNEYSEN, Hilbert v.: Surface Conductance Induced Dielectrophoresis of Semiconducting Single-Walled Carbon Nanotubes. In: *Nano Letters* 4 (2004), Nr. 8, 1395–1399. http://dx.doi.org/10.1021/nl0493794. – DOI 10.1021/nl0493794

[77] KRUPKE, Ralph ; HENNRICH, Frank ; LOEHNEYSEN, Hilbert v ; KAPPES, Manfred M.: Separation of metallic from semiconducting single-walled carbon nanotubes. In: *Science* 301 (2003), Nr. 5631, S. 344–347

[78] KUMAR, S. ; PIMPARKAR, N. ; MURTHY, J. Y. ; ALAM, M. A.: Theory of transfer characteristics of nanotube network transistors. In: *Applied Physics Letters* 88 (2006), Nr. 12, 123505. http://dx.doi.org/10.1063/1.2187401. – DOI 10.1063/1.2187401. – ISSN 00036951

[79] LÜDER, Ernst: *Bau hybrider Mikroschaltungen.* Berlin, Heiderlberg, New York : Springer-Verlag, 1977. – ISBN 3–540–08289–1

[80] LEE, Hang W. ; YOON, Yeohoon ; PARK, Steve ; OH, Joon H. ; HONG, Sanghyun ; LIYANAGE, Luckshitha S. ; WANG, Huiliang ; MORISHITA, Satoshi ; PATIL, Nishant ; PARK, Young J. ; PARK, Jong J. ; SPAKOWITZ, Andrew ; GALLI, Giulia ; GYGI, Francois ; WONG, Philip H.-S. ; TOK, Jeffrey B.-H. ; KIM, Jong M. ; BAO, Zhenan: Selective dispersion of high purity semiconducting single-walled carbon nanotubes with regioregular poly(3-alkylthiophene)s. In: *Nature Communications* 2 (2011), November, 541. http://dx.doi.org/10.1038/ncomms1545. – DOI 10.1038/ncomms1545

[81] LI, Jiantong ; ZHANG, Zhi-Bin ; ZHANG, Shi-Li: Percolation in random networks of heterogeneous nanotubes. In: *Applied Physics Letters* 91 (2007), December, Nr. 25, 253127–3. http://dx.doi.org/10.1063/1.2827577. – DOI 10.1063/1.2827577

[82] LI, Xiaokai ; GUARD, Louise M. ; JIANG, Jie ; SAKIMOTO, Kelsey ; HUANG, Jing-Shun ; WU, Jianguo ; LI, Jinyang ; YU, Lianqing ; POKHREL, Ravi ; BRUDVIG, Gary W. ; ISMAIL-BEIGI, Sohrab ; HAZARI, Nilay ; TAYLOR, André D.: Controlled Doping of Carbon Nanotubes with Metallocenes for Application in Hybrid Carbon Nanotube/Si Solar Cells. In: *Nano Letters* 14 (2014), June, Nr. 6, 3388–3394. http://dx.doi.org/10.1021/nl500894h. – DOI 10.1021/nl500894h. – ISSN 1530–6984

[83] LI, Yiming ; MANN, David ; ROLANDI, Marco ; KIM, Woong ; URAL, Ant ; HUNG, Steven ; JAVEY, Ali ; CAO, Jien ; WANG, Dunwei ; YENILMEZ, Erhan ; WANG, Qian ; GIBBONS, James F. ; NISHI, Yoshio ; DAI, Hongjie: Preferential Growth of Semiconducting Single-Walled Carbon Nanotubes by a Plasma Enhanced CVD Method. In: *Nano Letters* 4 (2004), February, Nr. 2, 317–321. http://dx.doi.org/10.1021/nl035097c. – DOI 10.1021/nl035097c

[84] LI, Yiming ; PENG, Shu ; MANN, David ; CAO, Jien ; TU, Ryan ; CHO, K. J. ; DAI, Hongjie: On the Origin of Preferential Growth of Semiconducting Single-Walled Carbon Nanotubes. In: *The Journal of Physical Chemistry B* 109 (2005), April, Nr. 15, 6968–6971. http://dx.doi.org/ 10.1021/jp050868h. – DOI 10.1021/jp050868h

[85] LINDNER, Albrecht: *Optimierung der Abscheidung von Kohlenstoffnanoröhren als halbleitende Schicht in Dünnschichttransistoren.* Stuttgart, Universität Stuttgart, Studienarbeit, August 2006

[86] LIU, Huaping ; TANAKA, Takeshi ; URABE, Yasuko ; KATAURA, Hiromichi: High-Efficiency Single-Chirality Separation of Carbon Nanotubes Using Temperature-Controlled Gel Chromatography. In: *Nano Letters* 13 (2013), Nr. 5, 1996–2003. http://dx.doi.org/10.1021/nl400128m. – DOI 10.1021/nl400128m. – ISSN 1530–6984

[87] LOLLI, Giulio ; ZHANG, Liang ; BALZANO, Leandro ; SAKULCHAICHAROEN, Nataphan ; TAN, Yongqiang ; RESASCO, Daniel E.: Tailoring (n,m) Structure of Single-Walled Carbon Nanotubes by Modifying Reaction Conditions and the Nature of the Support of CoMo Catalysts. In: *The Journal of Physical Chemistry B* 110 (2006), February, Nr. 5, 2108–2115. http://dx.doi.org/10. 1021/jp056095e. – DOI 10.1021/jp056095e

[88] MARUYAMA, Shigeo: *Shigeo Maruyama's Carbon Nanotube and Fullerene site.* http://www. photon.t.u-tokyo.ac.jp/~maruyama/nanotube.html. Version: 2010

[89] MEITL, Matthew A. ; ZHOU, Yangxin ; GAUR, Anshu ; JEON, Seokwoo ; USREY, Monica L. ; STRANO, Michael S. ; ROGERS, John A.: Solution casting and transfer printing single-walled carbon nanotube films. In: *Nano Letters* 4 (2004), Nr. 9, S. 1643–1647

[90] MIRRI, Francesca ; MA, Anson W. K. ; HSU, Tienyi T. ; BEHABTU, Natnael ; EICHMANN, Shannon L. ; YOUNG, Colin C. ; TSENTALOVICH, Dmitri E. ; PASQUALI, Matteo: High-Performance Carbon Nanotube Transparent Conductive Films by Scalable Dip Coating. In: *ACS Nano* 6 (2012), Nr. 11, 9737–9744. http://dx.doi.org/10.1021/nn303201g. – DOI 10.1021/nn303201g. – ISSN 1936–0851

[91] MISTRY, Kevin S. ; LARSEN, Brian A. ; BLACKBURN, Jeffrey L.: High-Yield Dispersions of Large-Diameter Semiconducting Single-Walled Carbon Nanotubes with Tunable Narrow Chirality Distributions. In: *ACS Nano* 7 (2013), Nr. 3, 2231–2239. http://dx.doi.org/10.1021/ nn305336x. – DOI 10.1021/nn305336x. – ISSN 1936–0851

[92] NIKOLAEV, Pavel ; BRONIKOWSKI, Michael J. ; BRADLEY, R. K. ; ROHMUND, Frank ; COLBERT, Daniel T. ; SMITH, K. A. ; SMALLEY, Richard E.: Gas-phase catalytic growth of single-walled carbon nanotubes from carbon monoxide. In: *Chemical Physics Letters* 313 (1999), November, Nr. 1-2, 91–97. http://dx.doi.org/10.1016/S0009-2614(99)01029-5. – DOI 10.1016/S0009–2614(99)01029–5. – ISSN 0009–2614

[93] NIRMALRAJ, Peter N. ; LYONS, Philip E. ; DE, Sukanta ; COLEMAN, Jonathan N. ; BOLAND, John J.: Electrical Connectivity in Single-Walled Carbon Nanotube Networks. In: *Nano Letters* 9 (2009), November, Nr. 11, 3890–3895. http://dx.doi.org/10.1021/nl9020914. – DOI 10.1021/nl9020914

[94] OPATKIEWICZ, Justin P. ; LEMIEUX, Melburne C. ; BAO, Zhenan: Influence of Electrostatic Interactions on Spin-Assembled Single-Walled Carbon Nanotube Networks on Amine-Functionalized Surfaces. In: *ACS Nano* 4 (2010), February, Nr. 2, 1167–1177. http://dx.doi.org/10. 1021/nn901388v. – DOI 10.1021/nn901388v

[95] ORTIZ-CONDE, A. ; GARCIA SANCHEZ, F. J. ; LIOU, J. J. ; CERDEIRA, A. ; ESTRADA, M. ; YUE, Y.: A review of recent MOSFET threshold voltage extraction methods. In: *Microelectronics Reliability* 42 (2002), April, Nr. 4–5, 583–596. http://dx.doi.org/10.1016/S0026-2714(02) 00027-6. – DOI 10.1016/S0026–2714(02)00027–6. – ISSN 0026–2714

[96] OSTFELD, Aminy E. ; CATHELINE, Amélie ; LIGSAY, Kathleen ; KIM, Kee-Chan ; CHEN, Zhihua ; FACCHETTI, Antonio ; FOGDEN, Siân ; ARIAS, Ana C.: Single-walled carbon nanotube transparent conductive films fabricated by reductive dissolution and spray coating for organic photovoltaics. In: *Applied Physics Letters* 105 (2014), December, Nr. 25, 253301. http://dx.doi.org/10. 1063/1.4904940. – DOI 10.1063/1.4904940. – ISSN 0003–6951, 1077–3118

[97] PARK, Tae-Jin ; BANERJEE, Sarbajit ; HEMRAJ-BENNY, Tirandai ; WONG, Stanislaus S.: Purification strategies and purity visualization techniques for single-walled carbon nanotubes. In: *Journal of Materials Chemistry* 16 (2006), Nr. 2, 141. http://dx.doi.org/10.1039/b510858f. – DOI 10.1039/b510858f. – ISSN 0959–9428

[98] PARK, Young-Bae ; HU, Liangbing ; GRÜNER, George ; IRVIN, Glen ; DRZAIC, Paul: Integration of Carbon Nanotube Transparent Electrodes into Display Applications. In: *SID Digest*, 2008, S. 537–540

[99] POPOV, Valentin N.: Carbon nanotubes: properties and application. In: *Materials Science and Engineering: R: Reports* 43 (2004), January, Nr. 3, 61–102. http://dx.doi.org/10.1016/j.mser.2003.10.001. – DOI 10.1016/j.mser.2003.10.001. – ISSN 0927–796X

[100] PRESTON, Colin ; HU, Liangbing: Silver Nanowires. Version: 2014. http://dx.doi.org/10.1007/978-3-642-35947-7_180-1. In: CHEN, Janglin (eds.) ; CRANTON, Wayne (eds.) ; FIHN, Mark (eds.): *Handbook of Visual Display Technology*. Berlin, Heidelberg : Springer Berlin Heidelberg, 2014. – DOI 10.1007/978–3–642–35947–7_180–1. – ISBN 978–3–642–35947–7, S. 1–14

[101] PROSS, Achim: *Transparente Kohlenstoff Nanoröhren Elektroden für Twisted Nematic Flüssigkristallzellen*. Stuttgart, Universität Stuttgart, Diplomarbeit, September 2007

[102] QU, Liangti ; DU, Feng ; DAI, Liming: Preferential Syntheses of Semiconducting Vertically Aligned Single-Walled Carbon Nanotubes for Direct Use in FETs. In: *Nano Letters* 8 (2008), September, Nr. 9, 2682–2687. http://dx.doi.org/10.1021/nl800967n. – DOI 10.1021/nl800967n. – ISSN 1530–6984

[103] REIBOLD, M. ; PAUFLER, P. ; LEVIN, A. A. ; KOCHMANN, W. ; PÄTZKE, N. ; MEYER, D. C.: Materials: Carbon nanotubes in an ancient Damascus sabre. In: *Nature* 444 (2006), November, Nr. 7117, 286. http://dx.doi.org/10.1038/444286a. – DOI 10.1038/444286a. – ISSN 1476–4687

[104] REICH, Dr. S. ; THOMSEN, Prof. C. ; MAULTZSCH, Dipl. Phys. J.: *Carbon Nanotubes - Basic Concepts and Physical Properties*. Weinheim : Wiley-VCH, 2003. – ISBN 3–527–40386–8

[105] REICH, S. ; MAULTZSCH, C. ; THOMSEN, C. ; ORDEJÓN, P.: Tight-binding description of graphene. In: *Physical Review B* 66 (2002), July, Nr. 3, 035412. http://dx.doi.org/10.1103/PhysRevB.66.035412. – DOI 10.1103/PhysRevB.66.035412

[106] REYNAUD, Olivier ; NASIBULIN, Albert G. ; ANISIMOV, Anton S. ; ANOSHKIN, Ilya V. ; JIANG, Hua ; KAUPPINEN, Esko I.: Aerosol feeding of catalyst precursor for CNT synthesis and highly conductive and transparent film fabrication. In: *Chemical Engineering Journal* 255 (2014), November, 134–140. http://dx.doi.org/10.1016/j.cej.2014.06.082. – DOI 10.1016/j.cej.2014.06.082. – ISSN 1385–8947

[107] ROBERTS, Mark E. ; LEMIEUX, Melburne C. ; SOKOLOV, Anatoliy N. ; BAO, Zhenan: Self-Sorted Nanotube Networks on Polymer Dielectrics for Low-Voltage Thin-Film Transistors. In: *Nano Letters* 9 (2009), July, Nr. 7, 2526–2531. http://dx.doi.org/10.1021/nl900287p. – DOI 10.1021/nl900287p

[108] SAITO, R ; DRESSELHAUS, G ; DRESSELHAUS, M. S.: *Physical properties of carbon nanotubes*. London : Imperial College Press, 1998. – ISBN 1–86094–093–5 978–1–86094–093–4 1–86094–223–7 978–1–86094–223–5

[109] SAITO, R. ; DRESSELHAUS, G. ; DRESSELHAUS, M. S.: Trigonal warping effect of carbon nanotubes. In: *Physical Review B* 61 (2000), January, Nr. 4, 2981–2990. http://dx.doi.org/10.1103/PhysRevB.61.2981. – DOI 10.1103/PhysRevB.61.2981

[110] SAITO, R. ; FUJITA, M. ; DRESSELHAUS, Gene ; DRESSELHAUS, Mildred S.: Electronic structure of graphene tubules based on C60. In: *Physical Review B* 46 (1992), Nr. 3, S. 1804–1811. – ISSN 0163–1829

[111] SANCHEZ-VALENCIA, Juan R. ; DIENEL, Thomas ; GRÖNING, Oliver ; SHORUBALKO, Ivan ; MUELLER, Andreas ; JANSEN, Martin ; AMSHAROV, Konstantin ; RUFFIEUX, Pascal ; FASEL, Roman: Controlled synthesis of single-chirality carbon nanotubes. In: *Nature* 512 (2014), August, Nr.

7512, 61–64. http://dx.doi.org/10.1038/nature13607. – DOI 10.1038/nature13607. – ISSN 0028–0836

[112] SARKAR, Jaydeep: *Sputtering Materials for VLSI and Thin Film Devices*. William Andrew, 2010. – ISBN 978–0–8155–1987–4. – Google-Books-ID: RVoKAgAAQBAJ

[113] SAVITZKY, Abraham. ; GOLAY, M. J. E.: Smoothing and Differentiation of Data by Simplified Least Squares Procedures. In: *Analytical Chemistry* 36 (1964), July, Nr. 8, 1627–1639. http://dx.doi.org/10.1021/ac60214a047. – DOI 10.1021/ac60214a047. – ISSN 0003–2700

[114] SCHADT, Martin ; HELFRICH, Wolfgang: Voltage-dependent optical activity of a twisted nematic liquid crystal. In: *Applied Physics Letters* 18 (1970), Nr. 4, S. 127–128

[115] SCHAU, Philipp: *Flexible Flüssigkristalldisplays mit transparenten Kohlenstoffnanoröhren-Elektroden*. Stuttgart, Universität Stuttgart, Diplomarbeit, July 2008

[116] SCHIESSL, Stefan P. ; FRÖHLICH, Nils ; HELD, Martin ; GANNOTT, Florentina ; SCHWEIGER, Manuel ; FORSTER, Michael ; SCHERF, Ullrich ; ZAUMSEIL, Jana: Polymer-Sorted Semiconducting Carbon Nanotube Networks for High-Performance Ambipolar Field-Effect Transistors. In: *ACS Applied Materials & Interfaces* (2014). http://dx.doi.org/10.1021/am506971b. – DOI 10.1021/am506971b. – ISSN 1944–8244

[117] SCHINDLER, Axel: *Untersuchung von Nanomaterialien als funktionale Schicht für flexible Displays*. Stuttgart, Universität Stuttgart, Diplomarbeit, July 2005

[118] SCHINDLER, Axel: Carbon Nanotube TFTs. Version: 2015. http://dx.doi.org/10.1007/978-3-642-35947-7_53-2. In: CHEN, Janglin (eds.) ; CRANTON, Wayne (eds.) ; FIHN, Mark (eds.): *Handbook of Visual Display Technology*. Springer Berlin Heidelberg, 2015. – DOI 10.1007/978–3–642–35947–7_53–2. – ISBN 978–3–642–35947–7, 1–33

[119] SCHINDLER, Axel ; LINDNER, Albrecht ; GOETTLING, Silke ; FRUEHAUF, Norbert ; NOVAK, James P. ; YANIV, Zvi: Solution-Deposited Carbon Nanotube Network TFTs on Glass and Flexible Substrates. In: *3rd International TFT Conference*. Rome, Italy, January 2007. – ISBN 1738–6047, S. 304–307

[120] SEIDEL, Robert ; GRAHAM, Andrew P. ; UNGER, Eugen ; DUESBERG, Georg S. ; LIEBAU, Maik ; STEINHOEGL, Werner ; KREUPL, Franz ; HOENLEIN, Wolfgang ; POMPE, Wolfgang: High-Current Nanotube Transistors. In: *Nano Letters* 4 (2004), May, Nr. 5, 831–834. http://dx.doi.org/10.1021/nl049776e. – DOI 10.1021/nl049776e

[121] SHERMAN, Robert: Carbon Dioxide Snow Cleaning. In: *Particulate Science and Technology* 25 (2007), January, Nr. 1, 37–57. http://dx.doi.org/10.1080/02726350601146424. – DOI 10.1080/02726350601146424. – ISSN 0272–6351

[122] SHERMAN, Robert ; HIRT, Drew ; VANE, Ronald: Surface cleaning with the carbon dioxide snow jet. In: *Journal of Vacuum Science & Technology A* 12 (1994), Nr. 4, S. 1876–1881. – ISSN 0734–2101. U.S. Copy

[123] SHIN, Dong H. ; KIM, Ji-Eun ; SHIM, Hyung C. ; SONG, Jin-Won ; YOON, Ju-Hyung ; KIM, Joon-dong ; JEONG, Sohee ; KANG, Junmo ; BAIK, Seunghyun ; HAN, Chang-Soo: Continuous Extraction of Highly Pure Metallic Single-Walled Carbon Nanotubes in a Microfluidic Channel. In: *Nano Letters* 8 (2008), December, Nr. 12, 4380–4385. http://dx.doi.org/10.1021/nl802237m. – DOI 10.1021/nl802237m

[124] SHULAKER, Max M. ; WEI, Hai ; PATIL, Nishant ; PROVINE, J. ; CHEN, Hong-Yu ; WONG, H.-S. P. ; MITRA, Subhasish: Linear Increases in Carbon Nanotube Density Through Multiple Transfer Technique. In: *Nano Letters* 11 (2011), Nr. 5, 1881–1886. http://dx.doi.org/10.1021/nl200063x. – DOI 10.1021/nl200063x. – ISSN 1530–6984

[125] SKÁKALOVÁ, Viera ; KAISER, A. B. ; WOO, Y.-S. ; ROTH, Siegmar: Electronic transport in carbon nanotubes: From individual nanotubes to thin and thick networks. In: *Physical Review B* 74 (2006), Nr. 8, 085403. http://dx.doi.org/10.1103/PhysRevB.74.085403. – DOI 10.1103/PhysRevB.74.085403

[126] SNOW, E. S. ; NOVAK, J. P. ; CAMPBELL, P. M. ; PARK, D.: Random networks of carbon nanotubes as an electronic material. In: *Applied Physics Letters* 82 (2003), March, Nr. 13, 2145–2147. http://dx.doi.org/10.1063/1.1564291. – DOI 10.1063/1.1564291

[127] STADERMANN, M. ; PAPADAKIS, S. J. ; FALVO, M. R. ; NOVAK, James P. ; SNOW, E. S. ; FU, Q. ; LIU, J. ; FRIDMAN, Y. ; BOLAND, J. J. ; SUPERFINE, R. ; WASHBURN, S.: Nanoscale study of conduction through carbon nanotube networks. In: *Physical Review B* 69 (2004), Nr. 20, S. 201402-1–3. – ISSN 0163-1829. U.S. Copy

[128] STOKES, Paul ; SILBAR, Eliot ; ZAYAS, Yashira M. ; KHONDAKER, Saiful I.: Solution processed large area field effect transistors from dielectrophoreticly aligned arrays of carbon nanotubes. In: *Applied Physics Letters* 94 (2009), Nr. 11, 113104. http://dx.doi.org/10.1063/1.3100197. – DOI 10.1063/1.3100197. – ISSN 00036951

[129] STRECKER, Michael: *Abscheidung von Kohlenstoffnanoröhren mittels Dielektrophorese.* Stuttgart, Universität Stuttgart, Diplomarbeit, October 2009

[130] THESS, Andreas ; LEE, Roland ; NIKOLAEV, Pavel ; DAI, Hongjie ; PETIT, Pierre ; ROBERT, Jerome ; XU, Chunhui ; LEE, Young H. ; KIM, Seong G. ; RINZLER, Andrew G. ; COLBERT, Daniel T. ; SCUSERIA, Gustavo E. ; TOMANEK, D. ; FISCHER, John E. ; SMALLEY, Richard E.: Crystalline Ropes of Metallic Carbon Nanotubes. In: *Science* 273 (1996), S. 483–487

[131] TOPINKA, Mark A. ; ROWELL, Michael W. ; GOLDHABER-GORDON, David ; MCGEHEE, Michael D. ; HECHT, David S. ; GRÜNER, George: Charge Transport in Interpenetrating Networks of Semiconducting and Metallic Carbon Nanotubes. In: *Nano Letters* 9 (2009), May, Nr. 5, 1866–1871. http://dx.doi.org/10.1021/nl803849e. – DOI 10.1021/nl803849e

[132] TOSHIMITSU, Fumiyuki ; NAKASHIMA, Naotoshi: Semiconducting single-walled carbon nanotubes sorting with a removable solubilizer based on dynamic supramolecular coordination chemistry. In: *Nature Communications* 5 (2014), October. http://dx.doi.org/10.1038/ncomms6041. – DOI 10.1038/ncomms6041

[133] TROTTIER, C. M. ; GLATKOWSKI, Paul ; WALLIS, Phillip ; LUO, J.: Properties and characterization of carbon-nanotube-based transparent conductive coating. In: *Journal of the SID* 13 (2005), Nr. 9, S. 759–763

[134] WANG, Yu ; LIU, Yunqi ; LI, Xianglong ; CAO, Lingchao ; WEI, Dacheng ; ZHANG, Hongliang ; SHI, Dachuan ; YU, Gui ; KAJIURA, Hisashi ; LI, Yongming: Direct Enrichment of Metallic Single-Walled Carbon Nanotubes Induced by the Different Molecular Composition of Monohydroxy Alcohol Homologues. In: *Small* 3 (2007), Nr. 9, 1486–1490. http://dx.doi.org/10.1002/smll.200700241. – DOI 10.1002/smll.200700241

[135] WANG, Yuhuang ; KIM, Myung J. ; SHAN, Hongwei ; KITTRELL, Carter ; FAN, Hua ; ERICSON, Lars M. ; HWANG, Wen-Fang ; AREPALLI, Sivaram ; HAUGE, Robert H. ; SMALLEY, Richard E.: Continued Growth of Single-Walled Carbon Nanotubes. In: *Nano Letters* 5 (2005), June, Nr. 6, 997–1002. http://dx.doi.org/10.1021/nl047851f. – DOI 10.1021/nl047851f

[136] WEISSTEIN, Eric W.: *Beta Distribution.* http://mathworld.wolfram.com/BetaDistribution.html

[137] WILLEY, Anthony D. ; HOLT, Josh M. ; LARSEN, Brian A. ; BLACKBURN, Jeffrey L. ; LIDDIARD, Steven ; ABBOTT, Jonathan ; COFFIN, Mallorie ; VANFLEET, Richard R. ; DAVIS, Robert C.: Thin films of carbon nanotubes via ultrasonic spraying of suspensions in N-methyl-2-pyrrolidone and N-cyclohexyl-2-pyrrolidone. In: *Journal of Vacuum Science & Technology B* 32 (2014), January, Nr. 1, 011218. http://dx.doi.org/10.1116/1.4861370. – DOI 10.1116/1.4861370. – ISSN 2166–2746, 2166–2754

[138] WONG, H.-S. P. ; AKINWANDE, Deji: *Carbon Nanotube and Graphene Device Physics.* Cambridge : Cambridge University Press, 2010 https://doi.org/10.1017/CBO9780511778124. – ISBN 978–0–511–77812–4

[139] XIE, Xu ; JIN, Sung H. ; WAHAB, Muhammad A. ; ISLAM, Ahmad E. ; ZHANG, Chenxi ; DU, Frank ; SEABRON, Eric ; LU, Tianjian ; DUNHAM, Simon N. ; CHEONG, Hou I. ; TU, Yen-Chu ; GUO, Zhilin ; CHUNG, Ha U. ; LI, Yuhang ; LIU, Yuhao ; LEE, Jong-Ho ; SONG, Jizhou ; HUANG, Yonggang ; ALAM, Muhammad A. ; WILSON, William L. ; ROGERS, John A.: Microwave purification of large-area horizontally aligned arrays of single-walled carbon nanotubes. In: *Nature*

Communications 5 (2014), November. http://dx.doi.org/10.1038/ncomms6332. – DOI 10.1038/ncomms6332

[140] XU, Hua ; ZHANG, Shixong ; ANLAGE, Steven M. ; HU, Liangbing ; GRÜNER, George: Frequency-and electric-field-dependent conductivity of single-walled carbon nanotube networks of varying density. In: *Physical Review B* 77 (2008), February, Nr. 7, 075418. http://dx.doi.org/10.1103/PhysRevB.77.075418. – DOI 10.1103/PhysRevB.77.075418

[141] YAN, Xing ; MONT, Frank W. ; POXSON, David J. ; SCHUBERT, Martin F. ; KIM, Jong K. ; CHO, Jaehee ; SCHUBERT, E. F.: Refractive Index Matched Indium Tin Oxide Electrodes for Liquid Crystal Displays. In: *Japanese Journal of Applied Physics* 48 (2009), 120203. http://dx.doi.org/10.1143/JJAP.48.120203. – DOI 10.1143/JJAP.48.120203. – ISSN 0021–4922

[142] YAN, Yehai ; CHAN-PARK, Mary B. ; ZHANG, Qing: Advances in Carbon-Nanotube Assembly. In: *Small* 3 (2007), Nr. 1, 24–42. http://dx.doi.org/10.1002/smll.200600354. – DOI 10.1002/smll.200600354

[143] YANG, Feng ; WANG, Xiao ; ZHANG, Daqi ; YANG, Juan ; LUO, Da ; XU, Ziwei ; WEI, Jiake ; WANG, Jian-Qiang ; XU, Zhi ; PENG, Fei ; LI, Xuemei ; LI, Ruoming ; LI, Yilun ; LI, Meihui ; BAI, Xuedong ; DING, Feng ; LI, Yan: Chirality-specific growth of single-walled carbon nanotubes on solid alloy catalysts. In: *Nature* 510 (2014), June, Nr. 7506, 522–524. http://dx.doi.org/10.1038/nature13434. – DOI 10.1038/nature13434. – ISSN 0028–0836

[144] YOON, Jinsu ; LEE, Dongil ; KIM, Chaewon ; LEE, Jieun ; CHOI, Bongsik ; KIM, Dong M. ; KIM, Dae H. ; LEE, Mijung ; CHOI, Yang-Kyu ; CHOI, Sung-Jin: Accurate extraction of mobility in carbon nanotube network transistors using C-V and I-V measurements. In: *Applied Physics Letters* 105 (2014), November, Nr. 21, 212103. http://dx.doi.org/10.1063/1.4902834. – DOI 10.1063/1.4902834. – ISSN 0003–6951, 1077–3118

[145] ZHANG, G. ; QI, P. ; WANG, X. ; LU, Y. ; LI, X. ; TU, R. ; BANGSARUNTIP, S. ; MANN, D. ; ZHANG, L. ; DAI, H.: Selective etching of metallic carbon nanotubes by gas-phase reaction. In: *Science* 314 (2006), Nr. 5801, S. 974–977

[146] ZHU, Lei: Modeling of a-Si:H TFT I-V Characteristics in the Forward Subthreshold Operation. (2005). https://uwspace.uwaterloo.ca/handle/10012/868

Curriculum vitae

August, 8 1977	Birth in Balingen, Germany
1988 - 1995	Grammar school Gymnasium Balingen, Germany
1995 - 1998	Technically oriented grammar school Technisches Gymnasium, Gewerbliche Schule Balingen, Germany
09/1998 – 10/1999	Rescue and ambulance service, in lieu of military service German Red Cross, Balingen, Germany
10/1999 – 07/2005	Studies of electrical engineering and information technology Specialization: Opto- and micro electronics Universität Stuttgart, Germany
04/2004 – 09/2004	Overseas internship Physical Optics Corporation, Electrooptics & Holography Division Torrance, California, USA
09/2005 – 12/2010	Research associate Institut für Systemtheorie und Bildschirmtechnik, Lehrstuhl für Bildschirmtechnik Universität Stuttgart, Germany
Since 01/2011	Project Manager at Swatch Group R&D - division Asulab Marin-Epagnier, Switzerland

www.ingramcontent.com/pod-product-compliance
Lightning Source LLC
Chambersburg PA
CBHW070730220326
41598CB00024BA/3382